IMPRESSUM/IMPRINT

Projektleitung/*Project Management:*
Sybilla Pütz

Gestaltung/*Design:*
Rena Chrysikopoulou, Susanne Schweikert
Reinzeichung/*Mechanicals:* Birgitta Sreball
CBCD Cool Blue Corporate Design GmbH,
Frankfurt am Main

Redaktion/*Editor:*
Marion Godau, Kleinmachnow

Lektorat/*Editorial Department:*
Claudia Ziesenhenne, SWB Communications,
Marbach am Neckar

Übersetzung/*Translations:*
Mark Robinson, Ilkley, GB

Litho/*Lithography:*
Hans Altenkirch Mediaproduktionen GmbH, Nieder-Olm

Druck/*Printing:*
Franz W. Wesel, Druckerei und Verlag GmbH & Co. KG,
Baden-Baden

Papier/*Paper:*
Römerturm Feinstpapier
Inhalt, Schutzumschlag/*Content, Dust Cover:*
Scheufelen Phoenix Imperial, halbmatt naturweiss 150 g/qm
Umschlag, Vor- und Nachsatz, Interview/
Cover, Pre- and Endpaper, Interview:
Zander Zeta gerippt, natur, 120 g/qm,

Die Deutsche Bibliothek –
CIP-Einheitsaufnahme
Tassilo von Grolman/
Hrsg.: Alex Buck.
[Übers.: Mark Robinson].–
Frankfurt am Main: Verl. form, 2000
(Designermonographien; 9)
ISBN 3-89802-021-5

Für meine liebe Frau Dagmar
For my dearest wife Dagmar

TASSILO VON GROLMAN

DESIGNERMONOGRAPHIEN/DESIGNER MONOGRAPHS 9

verlag form ●

6 Alex Buck: Vorwort/*Preface*

8 Marion Godau: Sinn und Sinnlichkeit/*Sense and Sensualilty*

20 Marion Godau: Innovation und Hitze/*Innovation and Heat*

38 Thomas Feicht: Naivität und Größe/*Naivety and Greatness*

48 Kurt Weidemann: Gestalt neu Denken/*Thinking Design Anew*

58 Hans-Ulrich von Mende: Kochlabor und Gaumenschmaus/
Cooking Laboratory and Feast for the Taste Buds

69 Interview mit Tassilo von Grolman:
Zwischen Emotion und Ratio/*Between Emotion and Reason*

82 Biographie/*Biography*

86 Renate Puvogel: Skulptur und Luxus/*Sculpture and Luxury*

96 Michael Schneider: Klarheit und Spiel/*Clarity and Play*

108 Marion Godau: Temperatur und Vielfalt/*Temperature and Variety*

120 Beatrix Freifrau von Wolff Metternich: Asche und Tee/*Ash and Tea*

130 Michael Conrad: Poesie und Welten/*Poetry and Worlds*

144 Mitarbeiter und externe Partner/*Staff and External Partners*

145 Kunden und Projekte/*Clients and Projects*

150 Autoren/*Authors*

152 Bibliographie und Fotonachweis/*Bibliography and Photo Credits*

VORWORT
PREFACE

Wer kennt sie nicht, Tassilo von Grolmans gläserne Teekanne, die sowohl funktional als auch ästhetisch hervorragend gelungen ist. In den 25 Jahren seiner Berufspraxis hat er ein beeindruckendes Œuvre geschaffen, vom Gaszähler bis zum Teddybär reicht sein Repertoire. Und damit nicht genug: Tassilo von Grolman überschreitet Grenzen. Gestalterische Kategorien wie Industrial Design, Corporate Design und Package Design sind an und für sich getrennt voneinander existierende gestalterische Welten, und doch gehören sie bei ihm zusammen und werden in seinem Werk mühelos und mit großer Kenntnis zusammengeführt. Tassilo von Grolman gestaltet Dinge für das Leben und aus dem Leben heraus. Diese haben so offensichtlich etwas mit unserem täglichen Leben zu tun, dass sie uns schon im Augenblick des ersten Kennenlernens vertraut erscheinen; vertraut, aber nicht langweilig, nicht üblich, nicht überflüssig, sondern freundlich, überraschend und unerwartet in den Kontext hineinpassend, in den sie von Tassilo von Grolman entworfen wurden. Die Gegenstände sind in ihren Funktionen nicht exaltiert und sie wirken auch nicht konstruiert, sondern sie sind notwendig und aus einem tiefem Verständnis für die Zusammenhänge unseres Alltags heraus entstanden. Den Alltag präzise und immer wieder neu zu begreifen, ihm Objekte anzubieten, die ihn bereichern, die ihn anschaulich machen, die Alltag erklären und ihn insgesamt in einen neuen Kontext stellen, das sind die Intentionen dieses Designers bei der Gestaltung von Gebrauchsgegenständen. Damit steht Tassilo von Grolman in einer Reihe mit großen deutschen Gestaltern wie zum Beispiel Wilhelm Wagenfeld oder auch Dieter Rams, die sich zutiefst um ein Verständnis des Wie des menschlichen Lebens im umbauten, wohnlichen Bereich bemüht haben. Wie diese hat auch Tassilo von Grolman Produkte geschaffen, die noch lange Zeit Bestand haben werden. Das vorliegende Buch ist der neunte Band der „Designermonographien“. Die Idee dieser Reihe ist, verschiedene gestalterische Positionen zeitgenössischer Gestalter zu beleuchten und damit unterschiedliche Schwerpunkte zu setzen. Der thematische Bogen der Reihe spannt sich deshalb auch nicht zufällig von Michele De Lucchi und Michael Graves über Alexander Neumeister und viele andere bis zu Tassilo von Grolman, dem der vorliegende Monographie-Band gewidmet ist. An dieser Stelle sei allen Akteuren und Helfern herzlich gedankt, die – im Vorder- wie im Hintergrund – zum Entstehen dieses Buches beigetragen haben.

Alex Buck

Who does not know Tassilo von Grolman's glass teapot, a teapot, which is an achievement in terms of both function and aesthetics. During a professional career of 25 years the designer created this and many other practical yet beautiful objects of everyday life. His repertoire extends from the gas meter to the teddy bear. And not satisfied with that: Tassilo von Grolman crosses borders. Creative categories such as industrial design, corporate design and packaging design usually exist in separate creative worlds, to him however they belong together and are effortlessly and with great knowledge brought together in his work. Tassilo von Grolman designs objects for life and from life. So obviously are these connected with our daily life that they seem familiar right from the first moment that we encounter them. Familiar but not boring, unusual, not superfluous but friendly, surprising and unexpectedly fitting into the context in which they have been designed by Tassilo von Grolman. The objects are not exaggerated in their functions nor do they seem at all artificial, but rather they are necessary, emerging from a deep understanding of the interrelationships of our daily life. To understand daily life precisely and always afresh, to make objects available which enrich it and make it come alive, to explain daily life and to place it altogether into a new context: these are the intentions of the Tassilo von Grolman's creations. In this way Tassilo von Grolman stands in a row of great German designers – such as Wilhelm Wagenfeld or Dieter Rams – who have been most concerned with understanding the needs of urban living and interior design on a human scale. Like them, Tassilo von Grolman has also created products that will long stand the test of time. This book is the ninth volume of the "Designers' Monographs". The idea of this series is to illuminate different creative positions of contemporary designers and thereby set different focal points. The thematic path of the series, therefore, does not by chance stretch from Michele De Lucchi and Michael Graves via Alexander Neumeister and many other designers to Tassilo von Grolman in this monographic volume. At this point all contributors and helpers are warmly thanked, who – at the forefront and in the background – have contributed to the production of this book.

Alex Buck

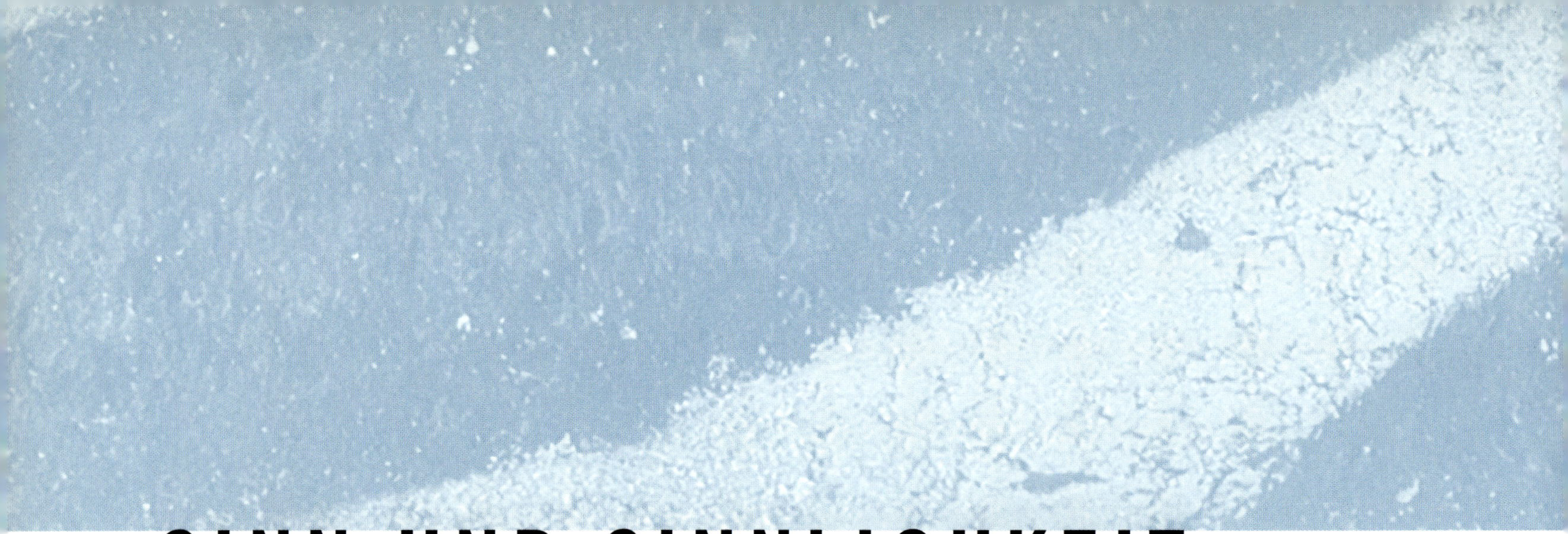

SINN UND SINNLICHKEIT

SPARGEL IN BRÖSEL-PETERSILIEN-BUTTER

1 Den Spargel schälen. Einen großen Topf halb mit Wasser füllen und zum Kochen bringen.

2 Dem Wasser Zucker, Salz und einen Schuss Wein oder den Saft der halben Zitrone zugeben. Den Spargel darin 15 Minuten kochen. Die Brötchen fein reiben.

3 In einem zweiten, kleinen Topf oder in einer Sauteuse die Butter schmelzen, die Brötchenbrösel zugeben, mit Salz und Pfeffer würzen und sehr langsam auf kleiner Flamme bräunen. Die Petersilie zufügen und einmal leicht umrühren.

4 Den Spargel über ein Tuch oder Küchenkrepp abtropfen und auf Tellern anrichten. Die Brötchenbrösel über den Spargel geben. Dazu reiche ich frische Pellkartoffeln und gekochten Schinken.

1 kg Stangenspargel

1 TL Zucker

Salz

trockener Weißwein

oder 1/2 Zitrone

2 alte Brötchen

200 g Butter

Pfeffer

1 EL fein gehackte Petersilie

MARION GODAU

SENSE AND SENSUALITY

ASPARAGUS IN PARSLEY BUTTER WITH BREADCRUMBS

1 kg asparagus spears
1 tsp sugar
Salt
Dry white wine or 1/2 lemon
2 old bread rolls
200 g butter
Pepper
1 tbsp finely chopped parsley

1 *Peel the asparagus. Half fill a large pan with water and bring to boil.*

2 *Add the sugar, salt and a dash of wine or the juice of half a lemon to the water. Boil the asparagus in the pan for 15 minutes. Finely grate the bread rolls.*

3 *Melt the butter in a second small pan or a sauteuse, add the breadcrumbs, season with salt and pepper and brown very slowly over a low heat. Add the parsley and gently stir once.*

4 *Drain the asparagus over a cloth or kitchen towel and serve onto plates. Pour the breadcrumb mixture over the asparagus. I serve peeled boiled potatoes and boiled ham with this dish.*

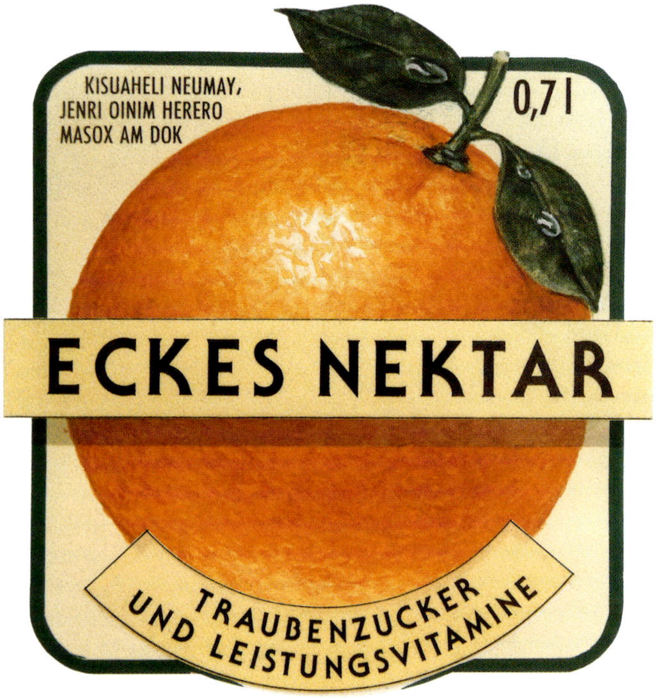

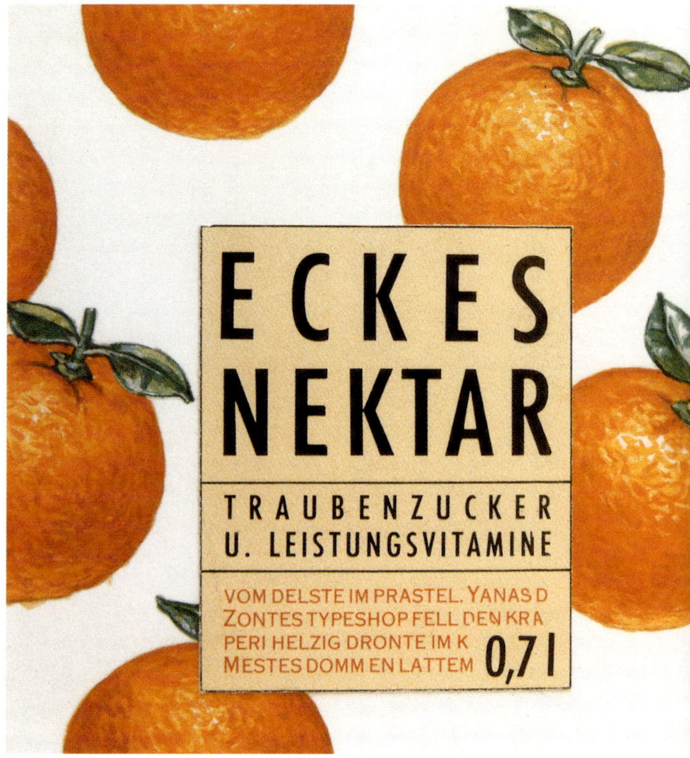

„ICH MUSS EINE TEEKANNE HABEN, DIE FUNKTIONIERT.“

Schreibtischartikel, Isolierkannen, Badaccessoires, Nudelverpackungen, Messestände, Buchgestaltungen – trotz der Bandbreite seiner Entwürfe wird Tassilo von Grolman meist mit einem einzigen Gegenstand verbunden: einer Teekanne, die der vielseitige Designer Anfang der achtziger Jahre kreierte. Von Grolman entwickelte eine elegante Variation des Teezubereitens, indem er ein Sieb entwickelte, das nahezu die ganze Kanne ausfüllt. Im Gegensatz zu herkömmlichen Kannen bot seine Lösung dem Tee damit endlich genügend Platz, sein Aroma zu entfalten. Der eigentliche Clou von „Mono Tea 1“ aber ist ihre ungewöhnliche Gestalt. Damit traf er den Nerv der Zeit und bereicherte Büro- wie Frühstückstische um eine vielschichtige Innovation. Für den halbkugelförmigen Kannenkörper wählte von Grolman das Material Glas und kombinierte es mit Sieb und Gestell aus Metall. Ergebnis ist eine „Teemaschine“, deren technoide Transparenz gleichzeitig filigrane Schwerelosigkeit und sinnliches Teeerlebnis verströmt. „Mono Tea 1“ ist längst ein Designklassiker, den man in zahlreichen Museen finden kann. Ihr Urheber ließ dem Erfolgsmodell noch weitere Varianten folgen. Gleichgültig ob es sich um Modelle „Filio“ „Tea Color“ oder „Chambord“ von Bodum handelt, von Grolmans gläserne Kannen kennt beinahe jeder Teetrinker, und zahlreiche Designer – leider auch Plagiatoren – wurden durch deren ungewöhnliche Ästhetik inspiriert.

"I MUST HAVE A TEAPOT THAT WORKS."

Articles for the office, insulating pots, bathroom accessories, packaging for pasta, exhibition stands, book designs – although the range of creations is wide, Tassilo von Grolman is associated with one particular object: a teapot, which the multifaceted designer created in the early eighties. Von Grolmann developed an elegant variation on the process for tea preparation by creating a tea strainer which almost filled the entire teapot. In contrast to conventional teapots, his solution finally allowed enough room for the flavour of the tea to develop. However, the real highlight of "Mono Tea 1" is its unusual shape. With this, he echoed the spirit of the times and enriched office and breakfast tables with a meaningful innovation. Von Grolman chose glass as the material for the hemispherical body of the teapot, and combined it with a metal strainer and stand. The result is a "tea machine", the technotransparency of which simultaneously exudes a filigree weightlessness and sensual tea experience. "Mono Tea 1" has long been regarded as a design classic, which can be found in numerous museums. Its creator provided for further variations to follow the successful original. Almost all tea drinkers know von Grolman's glass teapots. Numerous designers – and unfortunately also plagiarists – have been inspired by the unusual aesthetics of the design, be it of the "Filio", "Tea Color" or"Chambord" or model.

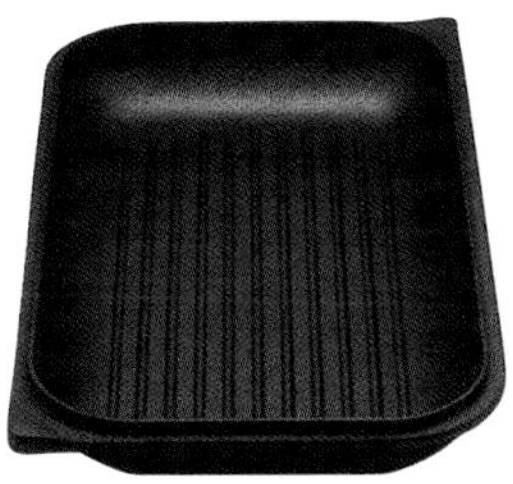
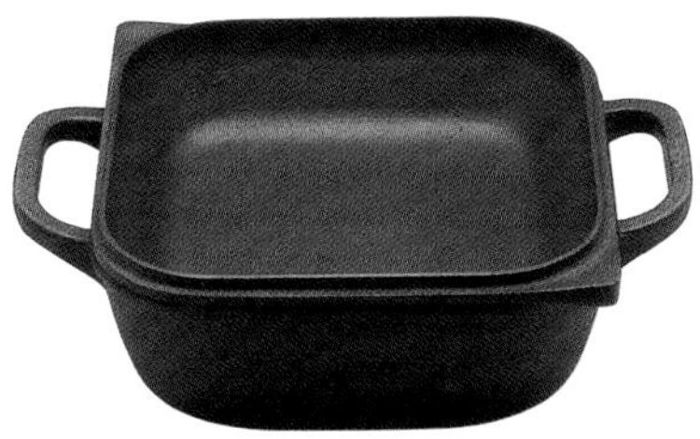
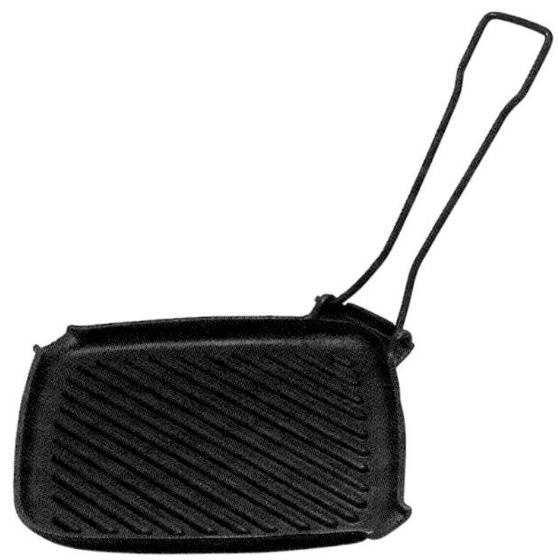

Linke Seite
Flaschenetiketten für Eckes. Entwurf in Zusammenarbeit mit Peter Engelhardt, 1986.

Rechte Seite
Der erste Auftrag: Aluminium-Guss-Serie für Berndes, 1975. Immer noch erfolgreich im Markt.

Left-hand side
Bottle labels for Eckes. Design in cooperation with Peter Engelhardt, 1986.

Right-hand side
The first commission: Aluminum-cast series for Berndes, 1975. Still successful in the market.

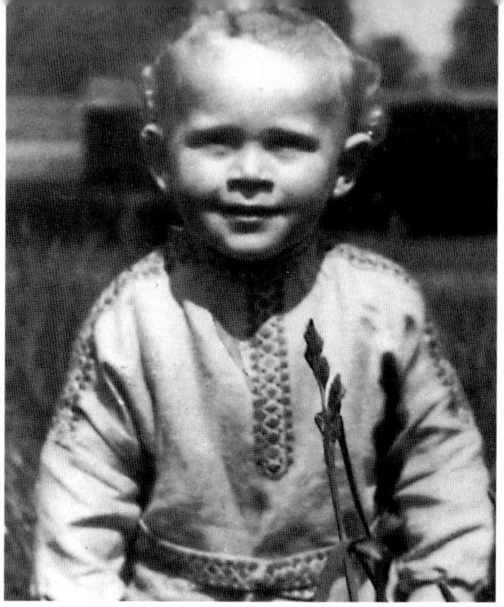

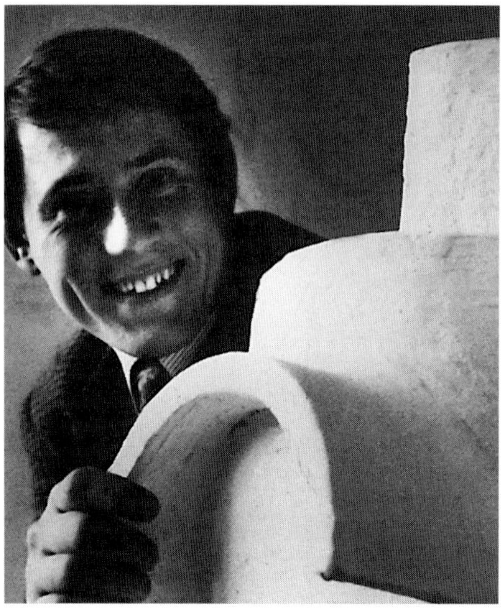

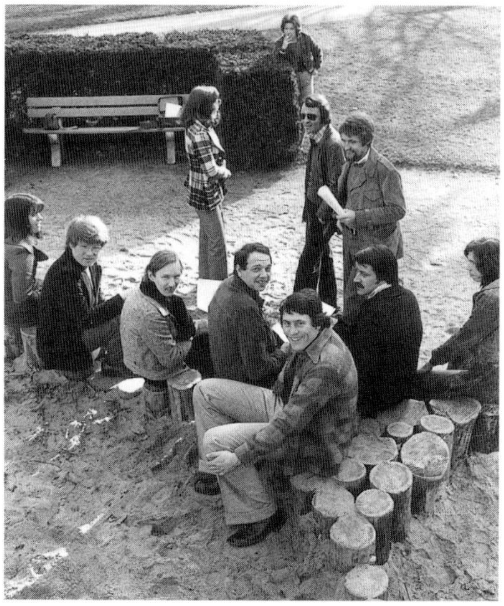

Oben
Tassilo von Grolman als Zweijähriger, 1944.

Links
Tassilo von Grolman während des Studiums, Werkkunstschule Kassel, Grundkurs „Plastische Durchdringung", 1969.

Rechts
Tassilo von Grolman mit VDID-Mitgliedern auf der Ronneburg/Frankfurt, 1974.

Above
Tassilo von Grolman at age two, 1944.

Left
Tassilo von Grolman during his studies, Werkkunstschule Kassel, basic course "Sculptural Penetration", 1969.

Right
Tassilo von Grolman with VDID-members on the Ronneburg (castle)/Frankfurt, 1974.

„EINEN ANDEREN BERUF WOLLTE ICH NIE."

Tassilo von Grolmans Weg zum Design verlief nicht ohne Hindernisse, vor allem weil der Spross aus altem Offiziersadel mit seinem Berufswunsch frontal gegen das Sicherheitsdenken seines Vaters prallte. Der hatte für seinen Sohn eine Karriere bei der Bundeswehr im Sinn und zwang ihm ein Maschinenbau-Studium auf. Mit 21 Jahren entzieht er sich dann auch kurz vor dem Maschinenbau-Examen dem väterlichen Druck. Er heuert auf dem Frachtschiff MS Santa Rosa an. Neun Monate lang ist er auf See. Die Route der Santa Rosa führt unter anderem an Nordamerikas Ostküste entlang, nach Australien, der Südsee und wieder zurück nordwärts nach Mittelamerika. Ständig wechselnde Eindrücke, Situationen und Mentalitäten lassen den Globetrotter im Zeitraffer reifen und prägen später seinen (Arbeits)Stil. Von Grolman ist heute in den unterschiedlichsten Disziplinen zu Hause. Ein Global Player mit Auftraggebern wie dem US-amerikanischen Tabakkonzern Philip Morris und der britischen Teacompany Twinings ist sein Haus auch privat keine abgeschottete Welt, sondern gleicht im Gegenteil einem Taubenschlag internationaler Gäste.

„ICH WAR EIN 68er."

Zurück in Deutschland zieht es von Grolman in die eingemauerte Metropole West-Berlin, das damals zudem ein Zentrum der Studentenbewegung war. Begeistert beteiligt er sich am Kampf gegen alte Zöpfe und fragwürdige Autoritäten. Gerade noch beendet er sein Maschinenbaustudium, danach gibt es kein Halten mehr: „Endlich machte ich Design, ich konnte mir gar nichts anderes vorstellen." Beginnend in Berlin ging er im geschichtsträchtigen Jahr 1968 nach Kassel, um sein zweites Studium so schnell wie möglich zu absolvieren. 10 Jahre nachdem von Grolman als Lehrling beschlossen hatte, Designer zu werden, war er endlich dort angekommen.

"I NEVER WANTED ANOTHER JOB."

As the offspring of old army aristocracy, Tassilo von Grolman's route to design was not without obstacle, largely because his chosen career path collided head-on with the protective intentions of his father. The latter had an army career in mind for his son and pushed him to study engineering. Then at the age of 21, just before taking his engineering exam, he escaped paternal pressure and enlisted for work on the freight ship, MS Santa Rosa. For nine months he went to sea. Amongst other places, the course of the Santa Rosa took him along the East Coast of North America, to Australia, the South Pacific and back north to Central America. Constantly changing impressions, situations and mentalities made the globetrotter mature as if in a time-lapse film, and later moulded the style of his work. Today, von Grolman is at home across a wide range of disciplines. He is a global player, whose clients include the US tobacco company, Philip Morris, and British tea company, Twinings. His home is not the world of a recluse, but more a busy meeting place for international guests.

"I WAS A PRODUCT OF 1968."

Back in Germany, von Grolman was drawn to the walled-in metropolis of West Berlin, the centre of the student movement at that time. Full of enthusiasm, he participated in the struggle against old customs and dubious authorities. When he completed his engineering studies, nothing more could stand in his way: "At last I was designing, I couldn't imagine doing anything else." After starting in Berlin, he went to Kassel in the historic year of 1968 to complete his second course of study as quickly as possible. 10 years after the decision made during his apprenticeship, von Grolman had finally become a designer.

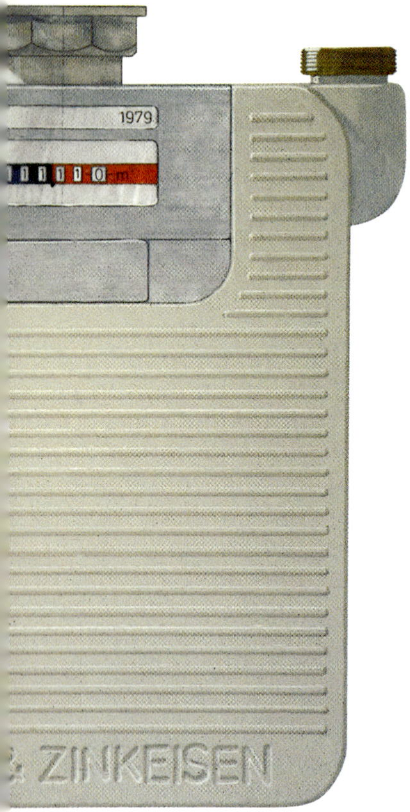

„DIE ORDNUNG DER DINGE IST KOLOSSAL WICHTIG."

Tassilo von Grolman gehört wie Alexander Neumeister, Jil Sander, Hartmut Esslinger oder Michael Erlhoff zu derjenigen Generation von Designern und Designtheoretikern, die in der Aufbruchstimmung des Nachkriegsdeutschlands aufwuchsen. Als Mitte der fünfziger Jahre die kleine Firma Braun mit ultrasachlich gestalteten Elektrogeräten reüssierte, war das für den 17 Jahre alten von Grolman gewissermaßen eine Initialzündung. Angeregt durch Brauns moderne Gestaltung entwarf der damalige Lehrling beim Elektrokonzern AEG einen Griff für den Staubsauger „Vampyrette" und reichte ihn als Verbesserungsvorschlag ein. Das brachte ihm prompt eine Prämie von DM 25,00 – damals für ihn immerhin ein Monatsgehalt. „Als ich das bekam, stand für mich fest: Ich werde Gestalter", sagt von Grolman rückblickend und ergänzt: „Ein Vorbild war immer Dieter Rams, der eine klare und konsequente Linie vertrat. Er hatte schon recht. Die Ordnung der Dinge ist kolossal wichtig." Auch Wilhelm Wagenfeld beeinflusste ihn stark, nicht zuletzt durch seinen Kasseler Professor Herbert Oestreich, einem Schüler des erklärten Industriegestalters. Folgerichtig sieht von Grolman sich nicht als Künstler, sondern als klassischen Produkt-Designer, dessen Augenmerk auf Vorgaben der Industrie wie Absatz und Fertigung liegen sollte.

„MICH INSPIRIERT DAS TÄGLICHE LEBEN."

Der heute nahe Frankfurt am Main lebende Designer hätte einst gern an der legendären Hochschule für Gestaltung Ulm studiert. Deren Ausrichtung auf Massenproduktion, Systemgedanke und Funktionalismus kamen ihm sehr entgegen. Mit Schlagwörtern wie „form follows emotion" kann er nichts anfangen: „Die Funktion ist nach wie vor für mich das Wichtigste überhaupt im Design. Ich will eine Teekanne haben, die funktioniert, die nicht kleckert. Das ist erst einmal die Voraussetzung. Dann kann ich sie sinnlich machen." Andererseits sagt von Grolman über sich: „Ich liebe eher das barocke Leben." Was zunächst widersprüchlich klingt, ist bei näherem Hinsehen ein Charakteristikum seiner Entwürfe. Denn trotz allem haftet ihnen oft etwas Spielerisches und Luxuriöses an, manchmal sogar Skulpturales (zum Beispiel seine Glasentwürfe für Lauscha). Dennoch: Was überwiegt, ist klare Formgebung, gute Handhabbarkeit und Liebe zum Detail. In der Summe macht gerade diese Spannung zwischen technischem Denken einerseits und sinnlicher Opulenz andererseits, zwischen Ratio und Phantasie von Grolmans Stil aus. Privat ist der vielseitige Designer ein passionierter Hobbykoch. Seine Lust an sinnlichen Stimulanzien ließ ihn zu einem Gourmet werden, der Freunde und Geschäftspartner begeistert bekocht. Weil Kulinarisches untrennbar mit dem Designer Tassilo von Grolman verbunden ist, findet sich in dieser Monografie erstmals eine Sammlung von Rezepten aus seinem reichen Erfahrungsschatz.

Linke Seite
Feuerfester Gaszähler, wiederverschließbar. 1979 für Dehm & Zinkeisen Schlumberger.

Rechte Seite
Etiketten für „Silkience"-Haarpflege von Gillette, 1979.

Flaschenkörper für die Waschlotion „Dermomild" für Merck AG, 1980.

Left-hand side
Fireproof gas meter, lockable, 1979 for Dehm & Zinkeisen Schlumberger.

Right-hand side
Labels for "Silkience" hair care products by Gillette, 1979.

Bottle shape for the skin cleansing lotion "Dermomild" for Merck AG, 1980.

"THE ORDER OF THINGS IS TREMENDOUSLY IMPORTANT."

Together with Alexander Neumeister, Jil Sander, Hartmut Esslinger or Michael Erlhoff, Tassilo von Grolman belongs to the generation of designers and design theorists who grew up in the atmosphere of the time of new departures in post-war Germany. The success achieved in the mid-fifties by the small company, Braun, with its ultra-functionally shaped electrical appliances, was, in a way, an explosive initiation for the seventeen-year-old von Grolman. Stimulated by Braun's modern design, the apprentice to the electrical company, AEG, developed a handle for the "Vampyrette" vacuum cleaner and presented it as a suggestion for improvement. This promptly brought him a bonus of 25 Marks – his monthly income at the time. "When I got that, it was clear: I will become a designer" says von Grolman looking back, and adds: "One of my role models was always Dieter Rams, who represented a clear and consistent line. He was quite right. The order of things is tremendously important." Wilhelm Wagenfeld was also a strong influence, not least through von Grolman's professor at Kassel, Herbert Oestreich, who had been a student of the acknowledged industrial designer. Consequently, von Grolman does not view himself as an artist, but as a classic product designer, whose attention should focus on the demands of industry such as sales and production.

"MY INSPIRATION IS EVERYDAY LIFE."

The designer, now resident near Frankfurt am Main, would have liked to study at the legendary School of Design in Ulm, being very comfortable with its orientation towards mass production, accommodating systems and functionalism. Catchphrases such as "form follows emotion" mean nothing to him: "Without exception, function is the most important aspect of design. I want a teapot that works, that does not spill. That is the first prerequisite. Then I can make it sensuous." On the other hand, von Grolman says, "I kind of prefer the baroque life." What initially sounds contradictory is, at a closer look, a characteristic of his designs. Over all, there is something playful and luxurious about them – sometimes even sculptural (his glass designs for Lauscha, for example). However, what dominate are clear design, good handling and love of detail. This tension between a technical approach and sensuous opulence on the one hand, and between reason and fantasy on the other, determines the totality of von Grolman's style. At home, the multifaceted designer is a passionate amateur cook. His enjoyment of sensual stimulation has led him to become a gourmet, enthusiastically cooking for friends and business partners. Because cooking is an intrinsic part of the designer von Grolman, this monograph includes, for the first time, a collection of recipes from his rich treasure chest of experience.

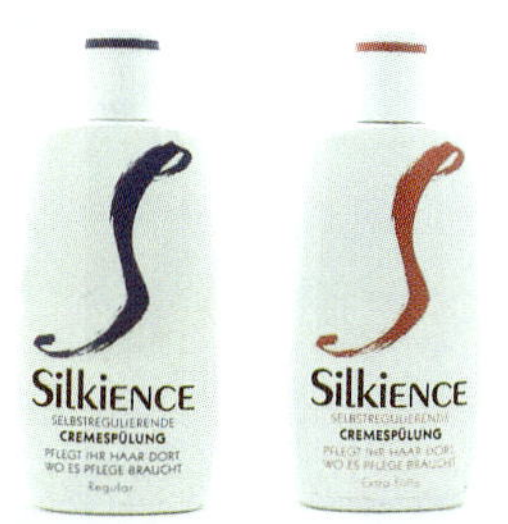

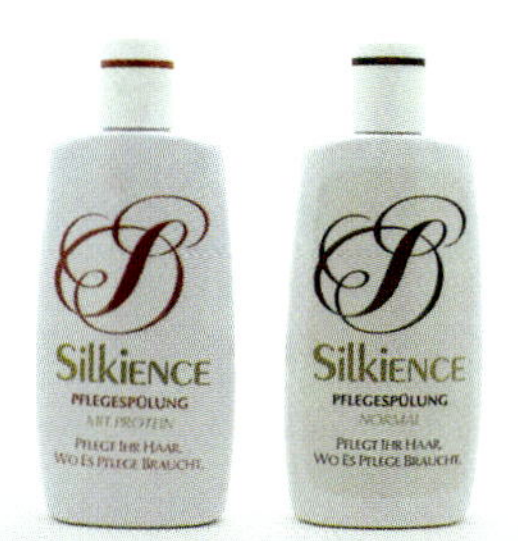

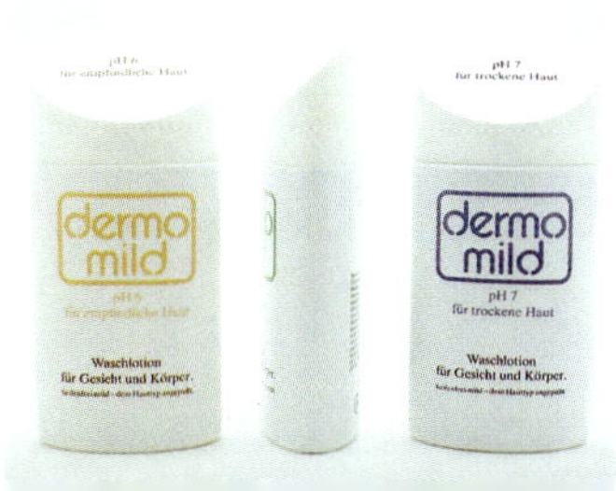

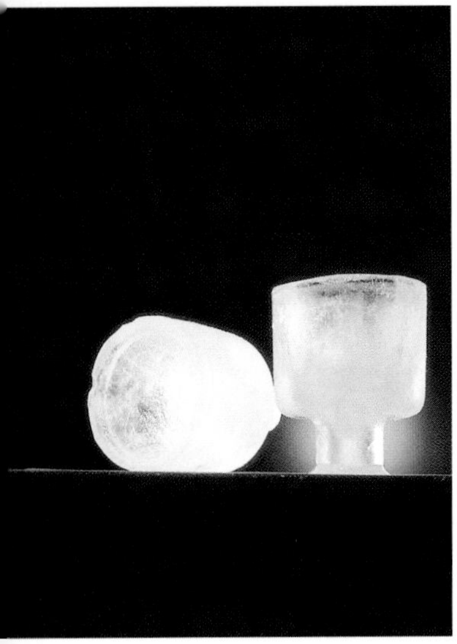

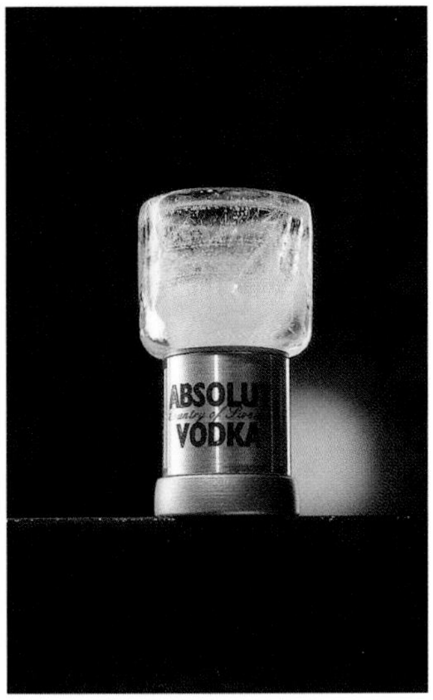

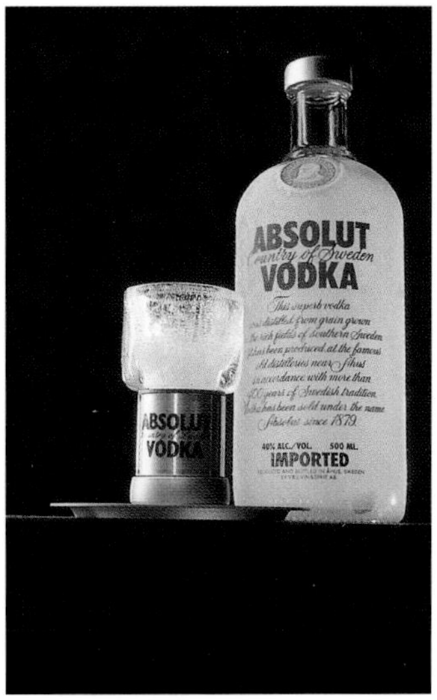

„ALLE DISZIPLINEN SIND MITEINANDER VERNETZT."

Als großen Glücksfall seiner Laufbahn sieht von Grolman seine Design-Ausbildung an der damals schon interdisziplinär ausgerichteten Werkkunstschule Kassel. Dort lernte er, „dass es außer Produktdesign die Malerei, die Architektur, die Typographie, natürlich auch die Bildhauerei, Textil und Mode gibt und dass all diese Disziplinen miteinander vernetzt sind." In jenem interdisziplinären Arbeiten sieht von Grolman eine der wichtigsten Voraussetzungen, um als Designer Erfolg zu haben. „Sie müssen zum Beispiel etwas von Typographie verstehen. Sie müssen kein Typograph sein, aber Sie müssen zumindestens verstehen, warum sie diese Typographie für dieses Feld bevorzugen statt eine andere." Noch heute arbeitet sein Büro in den verschiedensten Designdisziplinen wie Produkt-, Grafik-, Interior- und Corporate Design.

„OHNE WERBUNG LÄUFT NICHTS."

Seine Designerkarriere begann Anfang der siebziger Jahre in der Frankfurter Werbeagentur TBWA, die Michael Conrad und Walter Lürzer damals gerade aufbauten. Für den frisch gebackenen Hochschulabsolventen von Grolman war diese Zeit von immenser Bedeutung. Zunächst für Verpackungen zuständig stellte er sehr schnell fest, dass Marketing enormen Einfluss auf Produkte, Design und Werbung hatte: „Wenn Sie ein neues Produkt auf den Markt bringen und Sie machen keine Werbung und kein richtiges Marketing dafür, dann ist ein Produkt zum Tode verurteilt und da kann es noch so gut sein." Sein Fazit: „Ohne Werbung läuft nichts." Von Grolman beklagt, dass nicht wenige Auftraggeber den Sinn raffinierter Verpackungen oder fotografischer Bildsprache in Zweifel ziehen. Umgekehrt versteht er die Abneigung vieler Designer gegen Werbung nicht. Im Mangel an Vermarktungsgespür und Eigenwerbung sieht von Grolman, ganz Standespolitiker, ein großes Manko seiner Profession. So wurmt ihn, dass ein klares Urheberrecht für Designberufe bis heute fehlt. Nicht zuletzt um eine machtvolle Designerlobby zu etablieren, gründete von Grolman zusammen mit Weggefährten den Deutschen Designer Club (DDC). Ebenso zäh wie er einst seinen Berufswunsch durchsetzte, kämpft der erfolgreiche Gestalter um die (rechtliche) Anerkennung von kreativen Designleistungen.

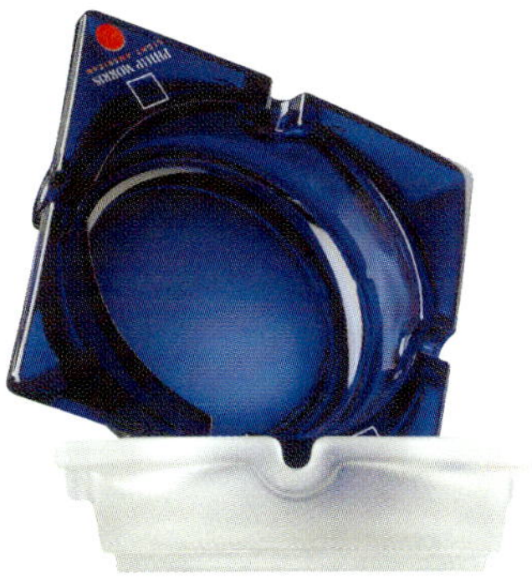

Linke Seite
Das Glas ist aus Eis, der Fuß nimmt die Form des Wodka-Flaschenverschlusses auf. Wettbewerbsarbeit für „Absolut Vodka", 1995.

Rechte Seite
Aschenbecher für die Gastronomie, Philip Morris Light American, 1996.

Left-hand side
The glass is made of ice, the foot repeats the shape of the vodka-bottle cap. Competition entry for "Absolut Vodka", 1995.

Right-hand side
Ashtray for gastronomy, Philip Morris Light American, 1996.

"ALL DISCIPLINES ARE INTERWOVEN."

Von Grolman views his design training at the, already at that time, interdisciplinary Kassel School of Arts and Crafts as a major stroke of luck in his life. There he learned that "apart from product design there is also painting, architecture, typography and, of course, sculpture, textiles and fashion – and that all these disciplines are interwoven. Von Grolman regards such interdisciplinary work as one of the most important prerequisites to being a successful designer. "For example, you have to understand something of typography. You don't have to be a typographer, but you must at least understand why you prefer a certain typography for an area to another." Today his company still works across different design disciplines such as product, graphic, interior and corporate design.

"NOTHING WORKS WITHOUT ADVERTISING."

His design career began in the early seventies in the Frankfurt advertising agency TBWA, which Michael Conrad and Walter Lürzer were just setting up. For von Grolman, a green and recent college graduate, this was an immensely important time. Initially responsible for packaging, he very quickly discovered that marketing had an enormous influence on product, design and advertising. "If you bring a new product onto the market and you have no advertising and no proper marketing for it, the product will face a death sentence, no matter how good it is." His conclusion: "Nothing works without advertising." Von Grolman laments that a not insignificant number of clients question the point of fancy packaging or the photographic communication of messages. On the other hand, he does not understand the aversion of many designers to advertising. Very much a politician for his profession, von Grolman sees a major shortcoming within his profession in the lack of commercial feel and self-promotion. Thus it rankles that there is still no clear copyright law for design professions today. Not least in order to establish a powerful design lobby, von Grolman and colleagues founded the German Designer Club (DDC). The successful designer fights just as hard for the (legal) recognition of creative services as he did to pursue his chosen career.

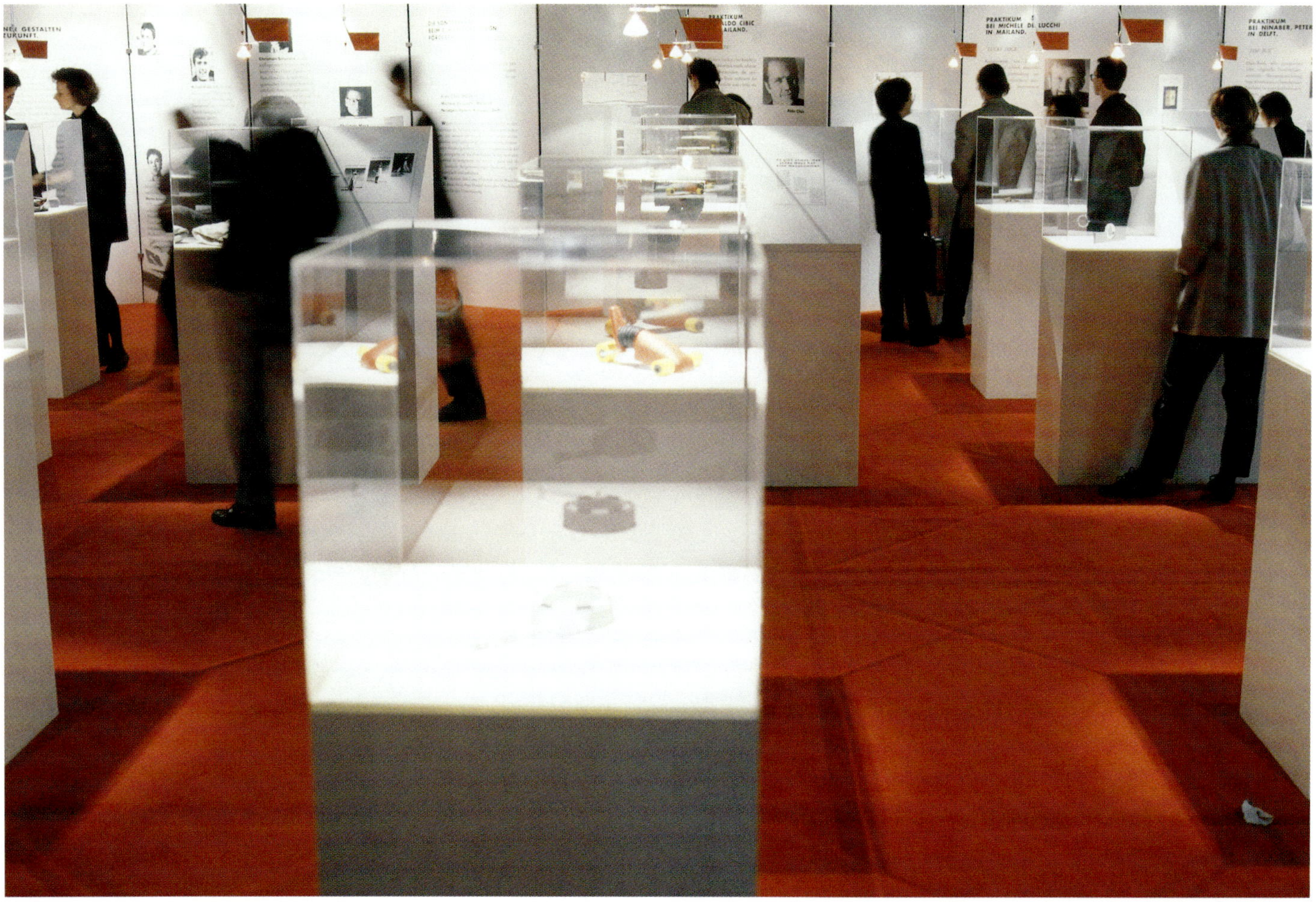

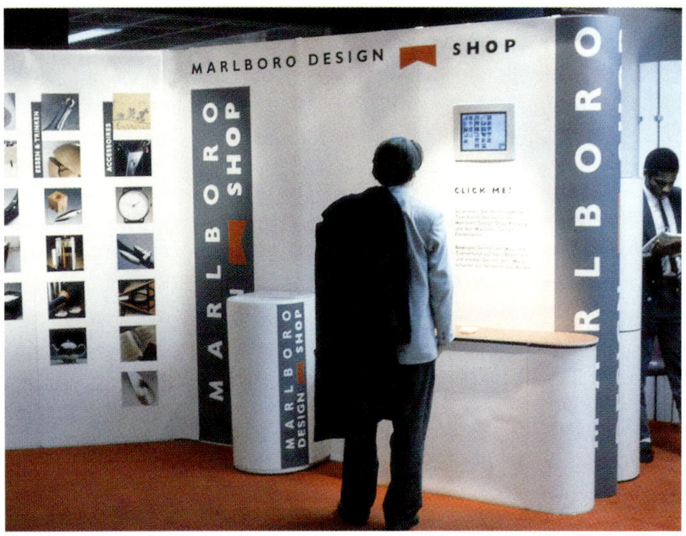

MARLBORO DESIGN SHOP
MARLBORO DESIGN SHOP
CLICK ME!
MARLBORO

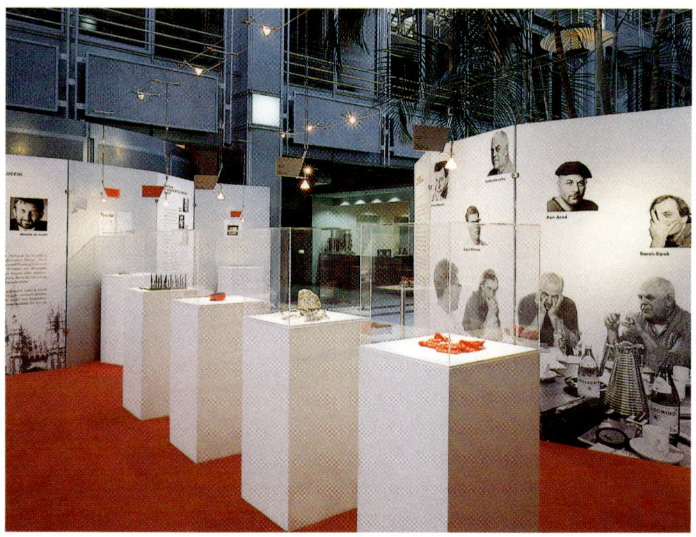

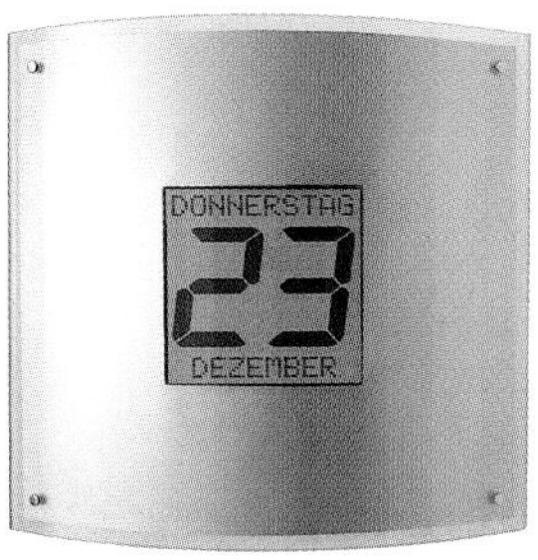

Linke Seite
Ausstellungs-Design „Marlboro Design Förderpreis“ von 1988 bis 1993.

Rechte Seite
Digitaler Wandkalender, Studie, 1993.

Left-hand side
Exhibition design "Marlboro Design Promotional Prize" from 1988 to 1993.

Right-hand side
Digital wall calendar, study, 1993.

„DEN DINGEN LEBEN EINHAUCHEN.“

Wenn man Tassilo von Grolmans facettenreiches Leben Revue passieren lässt, waren seine Lehre, seine Zeit zur See, sein Designstudium in Kassel und die Zeit in der Werbeagentur TBWA beziehungsweise bei Lürzer, Conrad sicherlich Dreh- und Angelpunkte, die sowohl den Designer als auch den Menschen von Grolman formten. Seine gestalterischen Wurzeln sieht der Weitgereiste vor allem im Funktionalismus von Wilhelm Wagenfeld und Dieter Rams, aber auch dem Bauhaus: „Das war lebensbejahend. Die Schule hat sich mit Farben beschäftigt, Werbung und Marketing gemacht und damit vieles vorweggenommen.“ Als Designer möchte von Grolman Dinge entwerfen, „die einfach noch nicht gelöst sind. Warum soll ich etwas tun, was andere ultimativ gelöst haben?“ Bedarf für Neues sieht er vor allem im Bad- und Sanitärbereich. Von Grolman hat jedoch für die organischen Formen des Franzosen Philippe Starck eine Schwäche ebenso wie für Alessandro Mendinis ironisch-verspielte Designzitate. „Wir brauchen mehr Humor, mehr Witz und mehr Frische im Design“, findet er und beklagt den biederen Ernst deutscher Couleur. So ist ihm, bei aller Betonung der Funktionalität, bloße Zweckerfüllung zu wenig. Vehement verweist er auf den emotionalen Aspekt eines Produktes. „Wenn man ein Produkt – ob es ein Möbelstück ist, ein Auto oder ein Besteck – sieht und dieses und kein anderes haben möchte, dann ist es gut. Dann ist die Beziehung zwischen Objekt und Mensch entfacht und wird auch Bestand haben. Ich will hinter die Dinge sehen und leblosen Objekten quasi Leben einhauchen. Ich glaube, Lust auf Sinnlichkeit macht Sinn – ganz stark in unserem Beruf.“

"BREATHE LIFE INTO THINGS."

On reviewing the many-facetted life of Tassilo von Grolman, it is apparent that it was his apprenticeship, his time at sea, his study of design in Kassel and his time at the advertising agency TBWA, Lürzer, Conrad, that were certainly pivotal and crucial in shaping the designer and also the man von Grolman. Above all, this citizen of the world sees his structural roots in the functionalism of Wilhelm Wagenfeld and Dieter Rams, and also in Bauhaus: "That was life-affirming. The school had concerned itself with colours, carried out some advertising and marketing and thereby pre-empted many things." As a designer, von Grolman would like to design things "for which a solution does not yet exist. Why should I do something that has ultimately been solved by other people?" Above all, he sees demand for innovations in the areas of bathroom and sanitary ware. However, von Grolman has a weakness for the organic shapes of the Frenchman Philippe Starck as well as for the ironically overplayed design statements of Alessandro Mendini. He feels that "we need more humour, more wit and more freshness in design" and bemoans the upright seriousness of the German mentality. Thus, for all his stressing of functionality, simply fulfilling the purpose is not enough for him. He vehemently points towards the emotional aspect of a product. "If you see a product – whether it is a piece of furniture, a car or cutlery – and you want to have it and no other – then it is good. Then the relationship between object and person is kindled and will further endure. I want to see behind things and to virtually revive lifeless objects. I believe that desire for the sensuous makes sense – and very strongly in our profession."

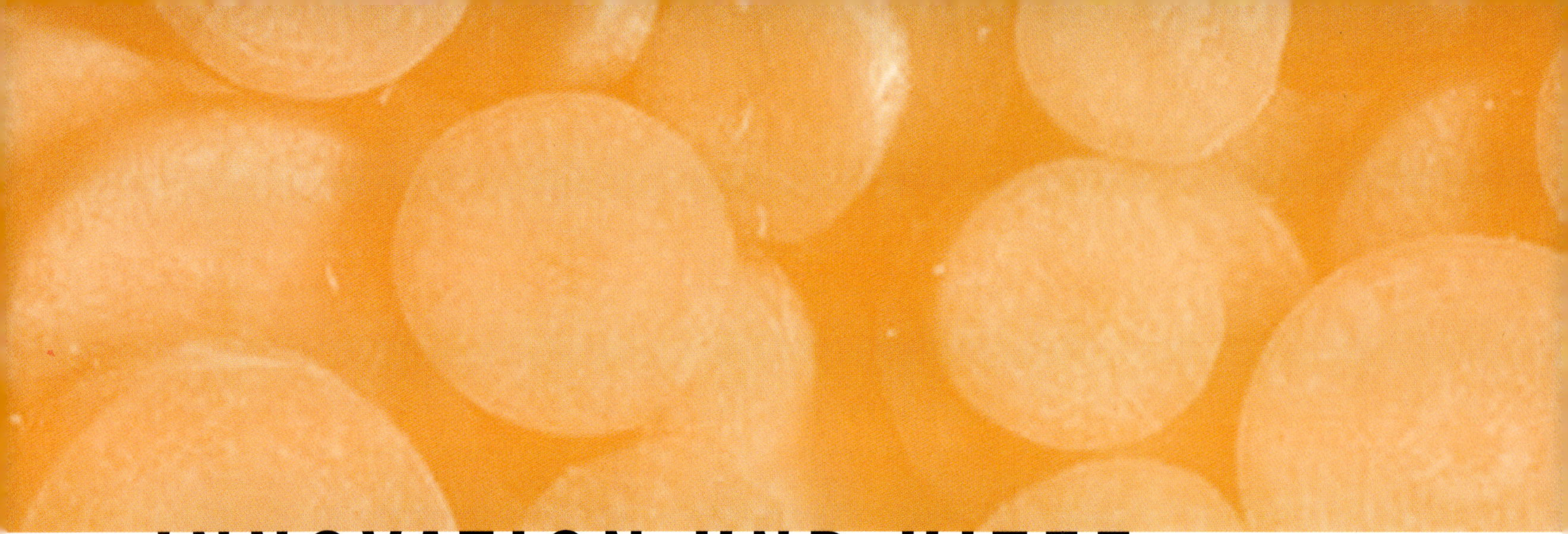

INNOVATION UND HITZE

200 g Karotten

5 Kartoffeln (normale Größe)

1 rote Paprikaschote

2 Zucchini

1/2 Staudensellerie

200 g Zuckererbsen

6 Tomaten

1 Stange Lauch

200 g Brokkoli

2 Gemüsezwiebeln

Basilikum

frischer Oregano

3 Knoblauchzehen

Muskatnuss

Kümmel

1 l Rinderbrühe

Salz, Pfeffer

MINESTRONE

1 Gemüse waschen, putzen und in kleine Würfel schneiden. Den Basilikum und den Oregano waschen und den Knoblauch schälen. Basilikum, Oregano und Knoblauch klein hacken, Muskat reiben.

2 Das Gemüse und den Kümmel in 1 l Rinderbrühe 10 Minuten kochen. Nach 8 Minuten die übrigen Gewürze hinzufügen. Abschmecken, eventuell noch einmal nachwürzen, da das Gemüse sehr viel Salz der Brühe aufsaugt. Nach weiteren 10 Minuten Garzeit ist die Minestrone servierbereit.

MARION GODAU

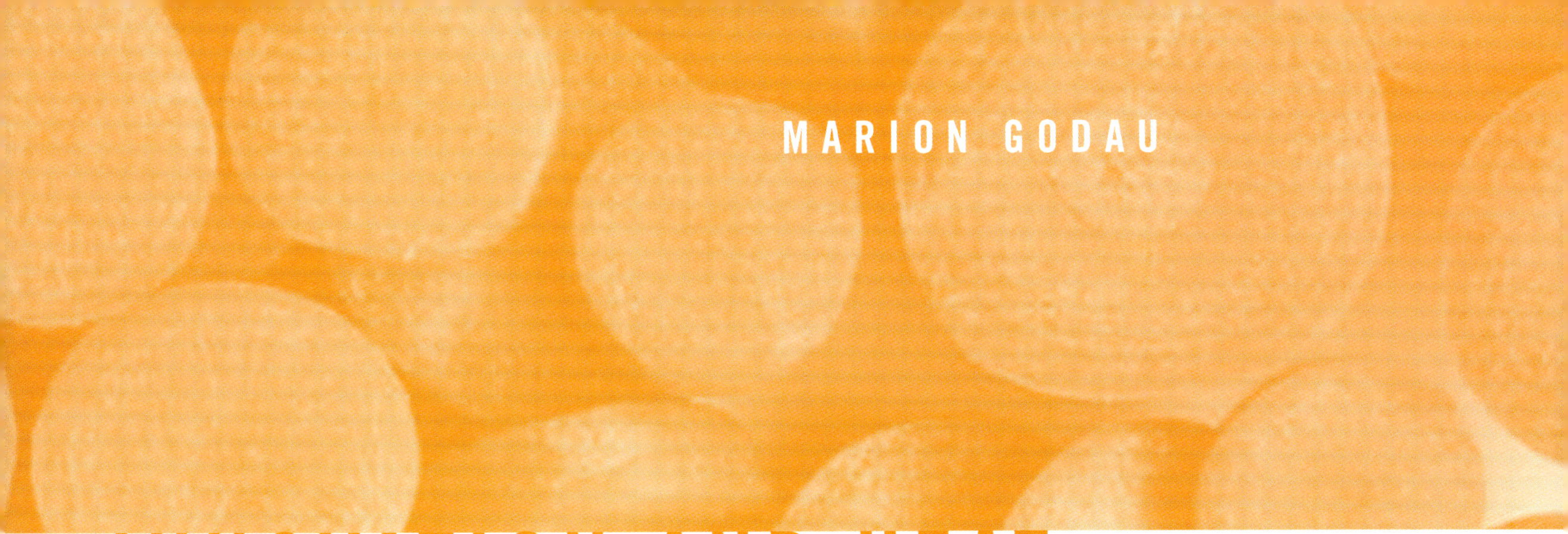

INNOVATION AND HEAT

200 g carrots
5 potatoes (normal size)
1 red pepper
2 courgettes
1/2 celery
200 g mange-tout peas
6 tomatoes
1 leek
200 g broccoli
2 onions
Basil
Fresh oregano
3 cloves of garlic
Nutmeg
Caraway seeds
1 l beef stock
Salt, pepper

MINESTRONE

1 *Clean, wash and finely dice the vegetables. Wash the basil and oregano and peel the garlic. Chop the basil, oregano and garlic finely. Grate the nutmeg.*

2 *Boil the vegetables and the caraway seeds in 1 litre of beef stock for 10 minutes. After 8 minutes, add the remaining herbs. Taste and season again as required as the vegetables soak up a lot of the salt in the stock. After a further 10 minutes cooking time, the minestrone is ready for serving.*

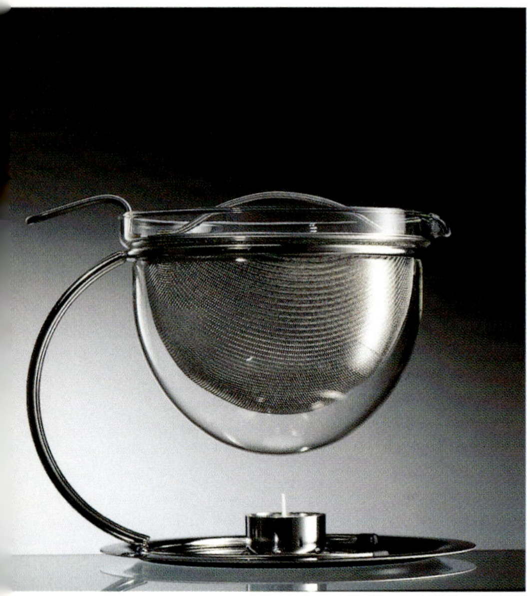

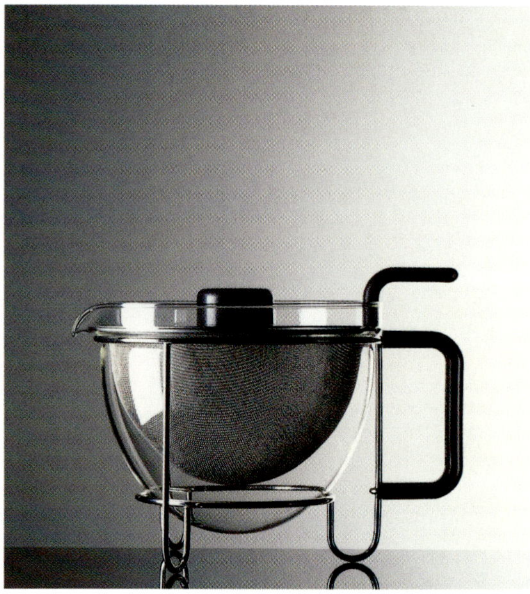

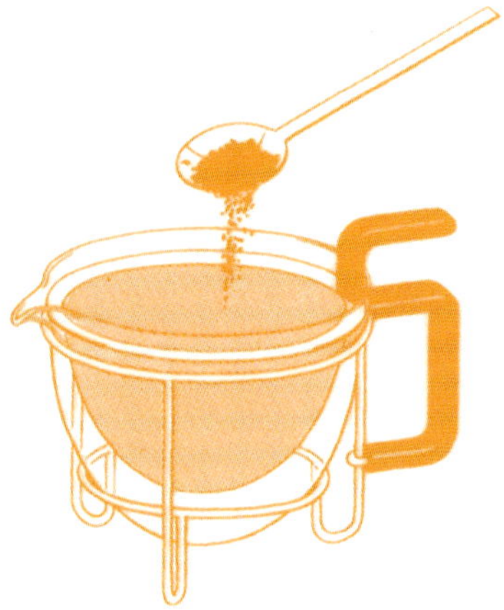

EIN EXPERIMENT

Bisher war ich mit meiner alten Teekanne aus Porzellan ganz zufrieden. Nun dieses Buch, und aus aktuellem Anlass steht Tassilo von Grolmans „Filio“ auf meinem Tisch. Brauche ich eine Teekanne aus Glas? Sind Bedienung und Zubereitungsverfahren besser als bei herkömmlichen Kannen? Ich entschließe mich zu einem Selbstversuch. Das ausladende Metallsieb hängt mit Teeblättern bestückt im Kannenkorpus und wartet auf siedendes Wasser. Beim Eingießen in die Kannenöffnung habe ich das etwas alberne Gefühl, ich füllte eine Glasschüssel mit kochendem Wasser. Die große Öffnung ist angenehm, aber ungewohnt. Langsam färbt sich das Wasser ein und bekommt seine schön anzusehende goldbraune Farbe, die in Porzellankannen buchstäblich im Dunkeln bleibt. Innerhalb von drei Minuten quellen die zuvor trockenen Krümel im Sieb auf und vergrößern ihr Volumen um ein Vielfaches. Einige Teeblattschnipsel steigen nach oben, andere sinken wieder nieder, sie bringen eine kreisförmige Bewegung in die nüchterne Boule. Ich beende das Schauspiel und nehme Metallsieb samt Blättern heraus. Zurück bleibt Tee, dessen warmer Ton einen reizvollen Kontrast zur technoiden Metall-Glas-Konstruktion der Kanne bildet. Ein Teelicht unter der gläsernen Halbkugel produziert interessante Lichtspiele. Passabel liegt „Filios“ metallener Kannenhenkel, zwei gebogene und parallel angebrachte Metallstangen, in der Hand. Das Ausgießen funktioniert tadellos, kein Kleckern, kein Tropfen, kein Schlabbern.

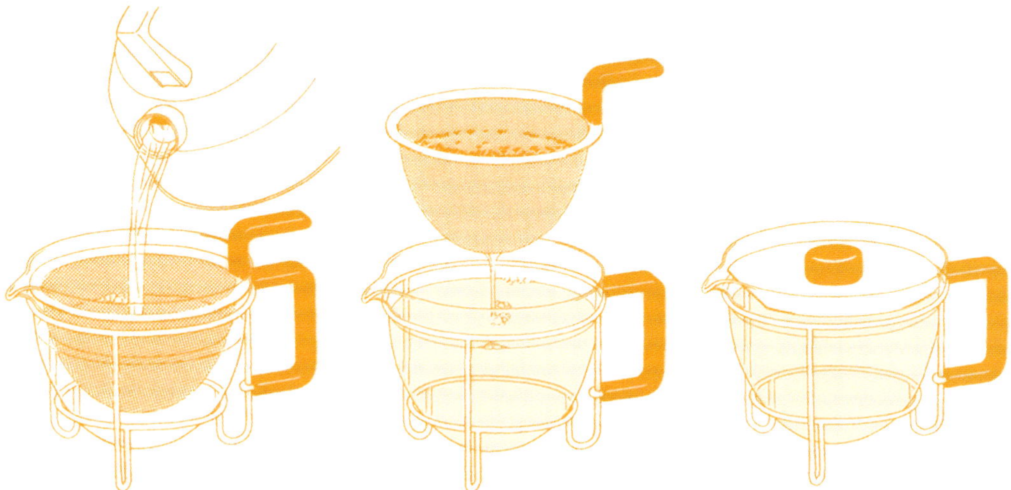

Linke Seite von links
„Mono filio" mit intergriertem Rechaud 1989, hervorgegangen aus der „Mono Filio" 1988.

„Mono 450" oder „Mono Tea 1", 1983, die erste in Serie gegangene Teekanne von Tassilo von Grolman für Mono Metallwarenfabrik GmbH.

„Mono filio 2", Tassilo von Grolman nennt sie „Katzenkanne".
Von all seinen Teekannen findet er sie am interessantesten.

Left-hand side, from left
"Mono Filio" with integrated réchaud, emerged from the "Mono Filio" 1988.

"Mono 450" or "Mono Tea 1", 1983, the first serially produced teapot by Tassilo von Grolman for Mono Metallwarenfabrik GmbH.

Mono Filio 2, Tassilo von Grolman calls it "Katzenkanne" (cat pot). Of all his teapots he finds this the most interesting.

AN EXPERIMENT

Previously I had been quite happy with my old porcelain teapot. Now with this book and current events, Tassilo von Grolman's "Filio" is on my table. Do I need a glass teapot? Does it work and make tea better than conventional teapots? I decide to try it for myself. The overhanging metal strainer filled with tea leaves is suspended in the body of the teapot, awaiting boiling water. On pouring the water into the teapot I have the somewhat foolish feeling of filling a glass bowl with boiling water. The large opening is pleasant but unfamiliar. Slowly the water colours and takes on that golden brown hue which is so pleasing to the eye and which is literally kept in the dark in porcelain teapots. Within three minutes, the previously dry crumbs in the strainer swell up and increase their volume many times over. Some scraps of tea rise to the top, others sink down again, bringing circular movement to the plain ball. I end the performance and take the metal strainer out, complete with leaves. Tea remains behind, its warm tones forming a charming contrast to the techno-metal-glass construction of the teapot. A candle underneath the glass hemisphere produces an interesting play on light. "Filios" metal teapot handle, made of two curved, parallel metal stems, moulds comfortably to the hand.

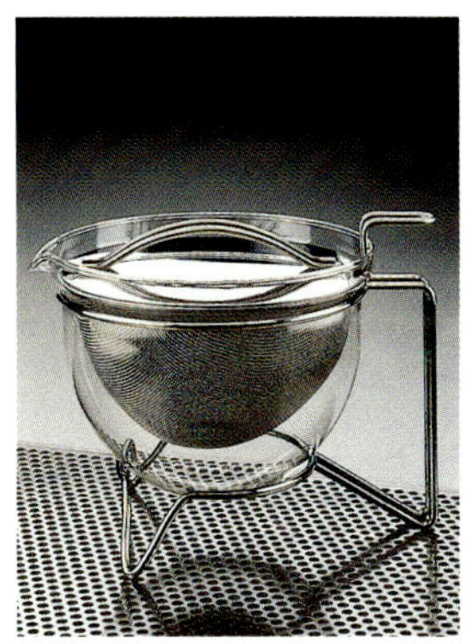

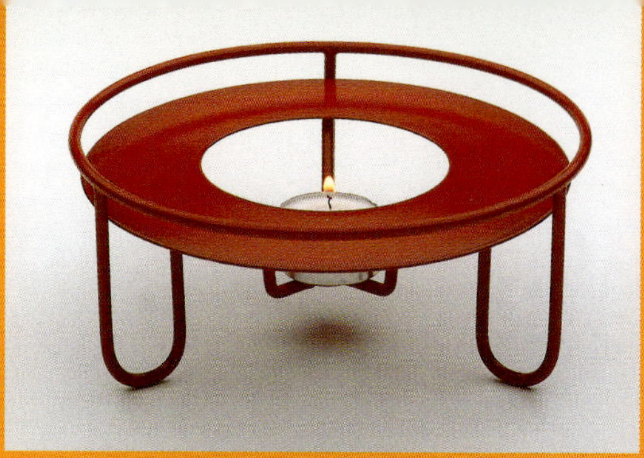

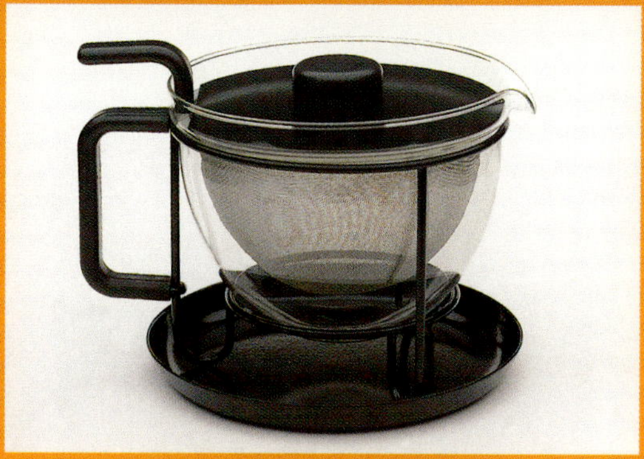

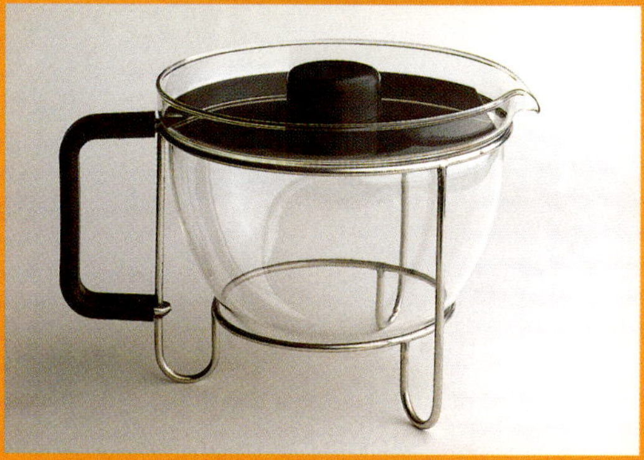

Linke Seite
Tassilo von Grolman schuf ein vielfältiges Mono Teekannen-Programm, indem er das Gestell in Form, Farbe und Material variierte und dazu verschiedene Stövchenversionen gestaltete.

Rechte Seite
„Mono Tea 1" mit Stövchen in Edelstahl und in den Farben Rot, Weiß und Schwarz, Hersteller Mono Metallwarenfabrik GmbH, 1982.

Left-hand side
Tassilo von Grolman created a versatile Mono tea pot program by varying the shape, colour and material of the frame and designing various types of réchaud.

Right-hand side
"Mono Tea 1" with réchaud in stainless steel and in the colours red, white and black, manufacturer Mono Metallwarenfabrik GmbH, 1982.

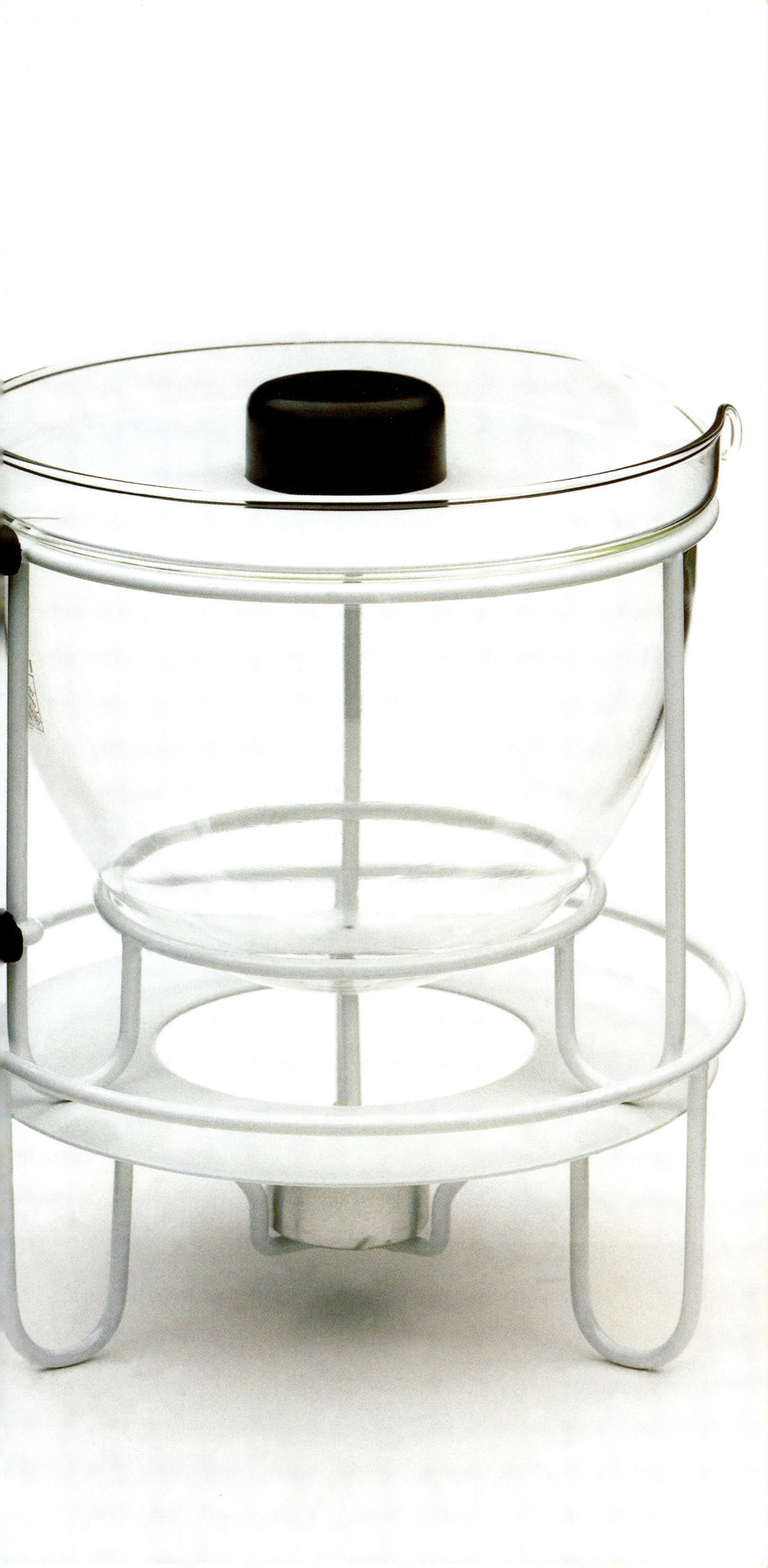

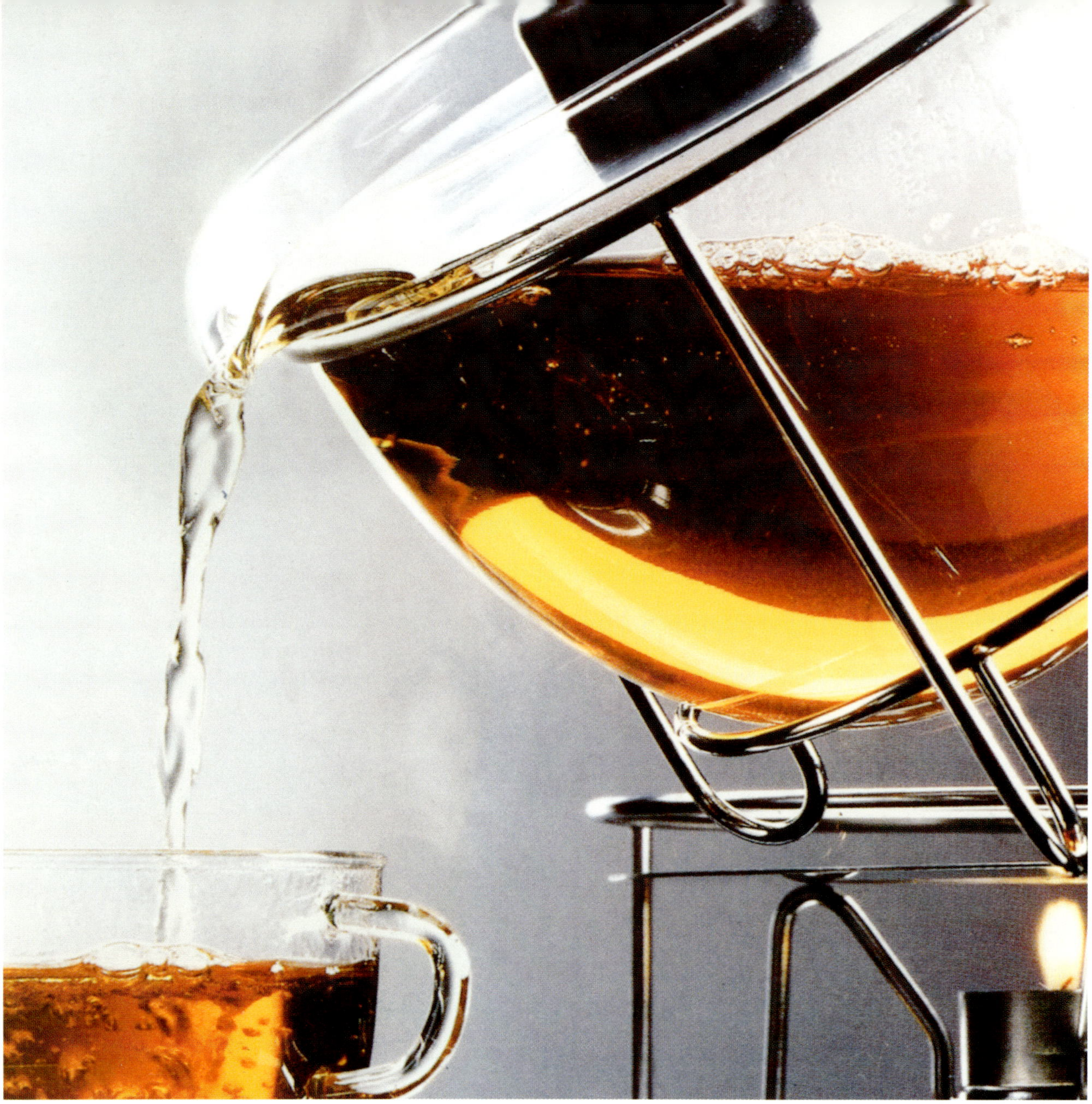

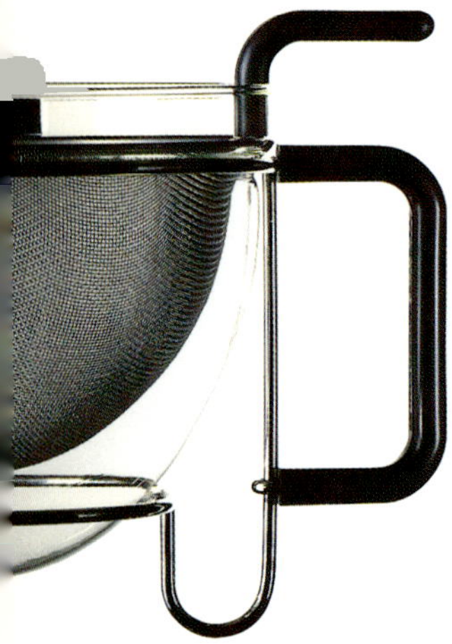

Auch der Tee schmeckt ausgezeichnet, und zufrieden betrachte ich die Teeboule, die elegant in ihrem Drahtgestell hängt und mich irgendwie an Mies van der Rohes Freischwinger mit seinen ausladenden Kufenbögen erinnert. Welch sinnlicher Zusatznutzen lässt sich doch aus dieser leicht zu bedienenden und simpel zu reinigenden Teekanne ziehen. Meine alte Kanne, die mir jahrelang treu ergeben war, steht etwas verloren auf dem Bord und wirkt plötzlich irgendwie behäbig. Ohne jeden Esprit. Schlechtes Gewissen lässt mich halbherzig in die Waagschale werfen, dass ja vielleicht die Teepatina ein Kritikpunkt wäre. Wahre Teegenießer nämlich schwören auf eine möglichst dicke Teestein-Kruste in der Kanne, die dem Tee fürderhin mehr Aroma geben soll. Und Tassilo von Grolmans Glaskannen mit stumpf dunkelbraun-fleckigem Korpus? Unmöglich. Allerdings weigern sich meine Geschmacksnerven, einen Unterschied zwischen bisherigem Teekochen mit Papier-Teefiltern plus Teepatina in der Kanne und von Grolmans Brühinnovation zu schmecken. Dennoch, auf die praktische Handhabung der Kanne möchte ich nicht mehr verzichten und schon gar nicht darauf, kontemplativ in den Tee zu schauen.

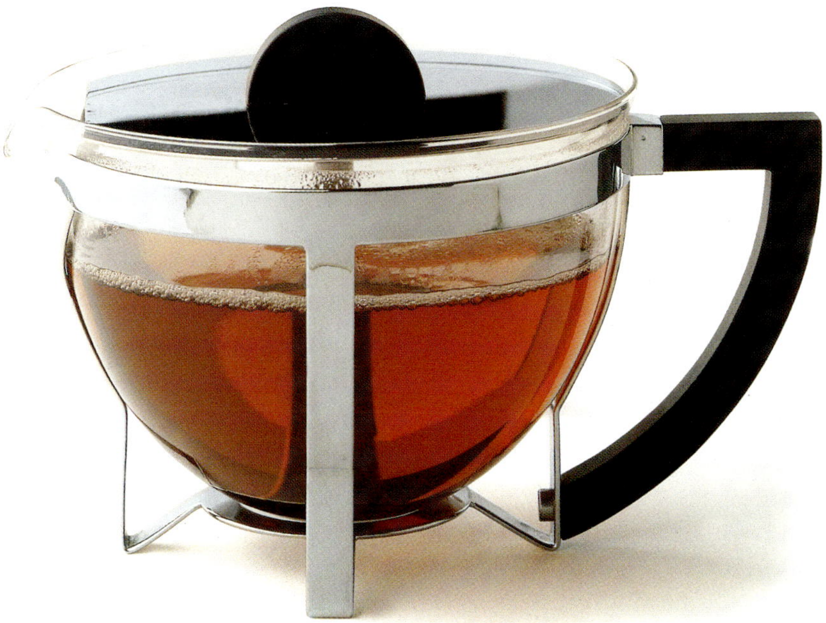

Linke Seite
Die „Mono Tea 1" wird auch „Mono 450" genannt.

Rechte Seite
„Chambord" war die Urform des Designklassikers und ging 1987 in Serie.

Left-hand side
The "Mono Tea 1" is also called "Mono 450".

Right-hand side
"Chambord" was the original form of the design classic and was serially produced starting 1987.

It pours perfectly: no spilling, no dripping, and no dribbling. The tea also tastes excellent and, satisfied, I regard the tea ball, which hangs elegantly in its wire stand and somehow reminds me of Mies van der Rohe's Free Swinger with its protruding running curves. What sensual additional purpose can be extracted from this easy to use and simple to clean teapot! My old teapot, which had remained faithful for years, stands rather forlorn on the shelf and suddenly seems somewhat stolid. Lacking in spirit. Triggered by my guilty conscience, I half-heartedly wonder whether the lack of tea residue may perhaps be a point for criticism of "Filio". Genuine tea connoisseurs actually swear by the thickest possible coating of old tea deposits in the teapot, which is thereafter believed to give the tea more flavour. And what of Tassilo von Grolman's glass teapots with a dull, dark brown stained body? Impossible. However my taste buds refuse to differentiate between teas brewed previously using tea filters plus residue and von Grolman's brewing innovation. That said, I would not like to deny myself the practical handling of the teapot and certainly not the reflective observation of the tea.

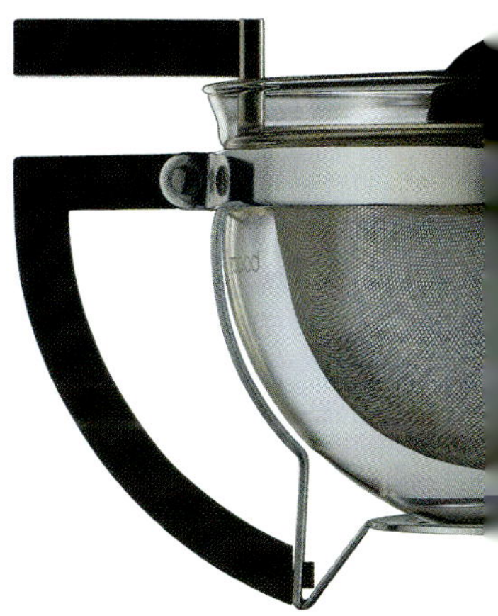

EIN NEUES MATERIAL

Gemessen an der über 2000 Jahre alten Geschichte des Teetrinkens ist die Idee, Glas als Material für Teeservice zu verwenden, noch sehr neu. Sie kam erstmals in den dreißiger Jahren auf. Ihr Urheber: Wilhelm Wagenfeld, einer der ersten Formgestalter Deutschlands. Das von ihm verwendete dünnwandige Glas verlieh dem bis dato klassischen Porzellan-, Metall- oder Steingutartikel eine ganz eigene, filigrane Wirkung. Seine Transparenz vermittelte fragile Schwerelosigkeit. Wagenfeld rückte mit Hilfe des Glases nicht die Ästhetik der Kanne in den Vordergrund, sondern die Ästhetik ihres Inhalts. Im Gegensatz zu von Grolmans gläsernen Kannenentwürfen wurde Wagenfelds Service jedoch mundgeblasen, die einzelnen Komponenten der Kanne wie Henkel und Tülle mussten handwerklich kompliziert miteinander verbunden werden. Ein halbes Jahrhundert später gab es immer noch wenige Glas-Teekannen auf dem Markt. Zwar hatte das schwedische Unternehmen Boda Nova sich seit Anfang der siebziger Jahre mit gläsernen Teekannen plus Stövchen hervorgetan, diese unterschieden sich aber in ihrer Form von keramischen Erzeugnissen wenig. Als von Grolman 1983 mit „Mono Tea 1" reüssierte, steckte sein Entwurf nicht nur voll funktionaler Innovationen, sondern definierte die Gestalt von Teekannen neu.

A NEW MATERIAL

In the context of the 2000-year history of tea drinking, the idea of using glass as the material for a tea service is still very new. It first appeared in the thirties. Its originator: Wilhelm Wagenfeld, one of the first designers of form in Germany. The thin-walled glass he used lent the classic porcelain, metal or earthenware articles of the time a quite unique filigree effect. Its transparency evoking fragile weightlessness. With the help of glass, Wagenfeld emphasized the aesthetics of the contents of the teapot over the aesthetics of the teapot itself. In contrast to von Grolman's glass teapot designs, Wagenfeld's were hand blown. The individual components of the teapot such as the handle and the spout had to be joined together by a complicated piece of craftsmanship. Half a century later, there were still only a few glass teapots on the market. Although the Swedish company, Boda Nova, had distinguished itself with the production of glass teapots with warmers since the early seventies, their design varied little from that of ceramic products. When von Grolman was successful in 1983 with "Mono Tea 1", his design was not only full of completely functional innovation but also redefined the shape of teapots.

Linke Seite
„Mono LAB 2" mit integriertem Rechaud, 1985.

Rechte Seite
„Mono LAB 2" mit charakteristischen Kunststoffelementen.

Left-hand side
"Mono LAB 2" with integrated réchaud, 1985.

Right-hand side
"Mono LAB 2" with its characteristic plastic elements.

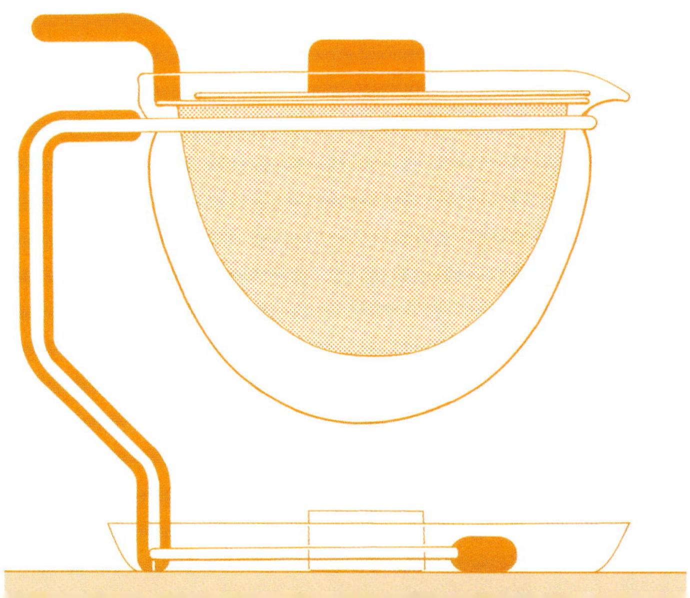

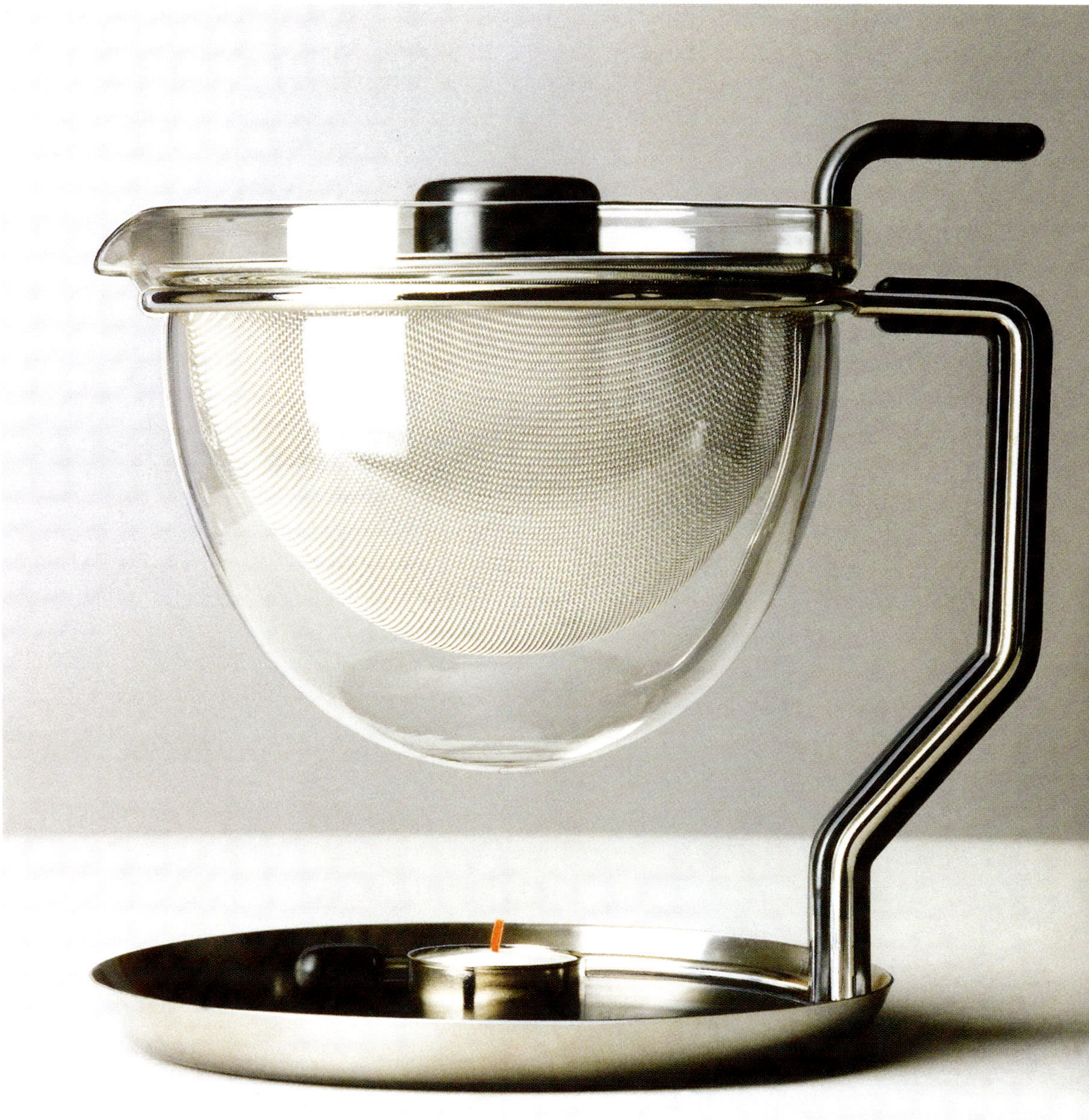

AUSLÖSER EIN VORTRAG

Auf seine Aufsehen erregende Teekannen-Idee kam Tassilo von Grolman während einer Veranstaltung im Rosenthal Studiohaus in Frankfurt. Jener Abend im Herbst 1979 drehte sich um das Thema Tee. Anbau, Verarbeitung, Zubereitung, Wirkung – Teeexperten beleuchteten das Thema von allen Seiten. Tassilo von Grolman, selbst passionierter Teetrinker, referierte über die Gestaltung von Teekannen und kam irritiert zu dem Schluss, dass man für den optimalen Teegeschmack eigentlich nach wie vor zwei Kannen braucht. Bei diesem uralten Verfahren wird der Tee in einer Kanne aufgebrüht und dann durch ein Sieb in eine zweite, vorgewärmte Kanne geseiht. Auch Teekannenklassiker wie die von Walter Gropius oder besagte gläserne Kanne von Wilhelm Wagenfeld waren so gedacht. Die winzigen Teesiebe sollten lediglich übrig gebliebene Blätter aus der ersten Kanne auffangen.

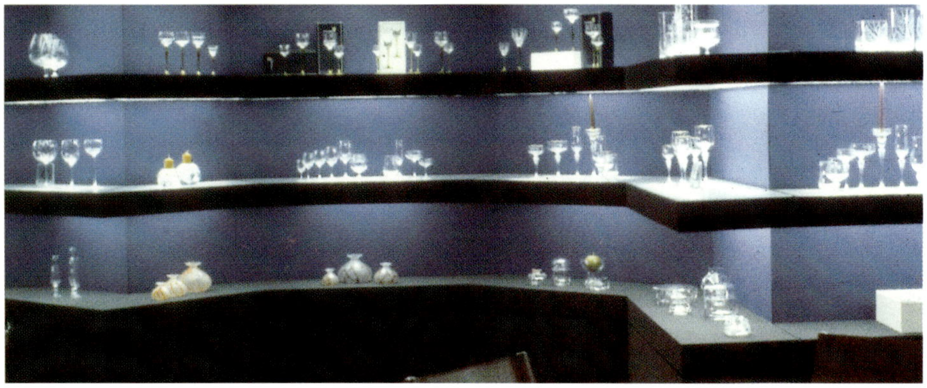

Das Gesagte revuepassierend beschloss von Grolman noch am selben Abend, eine Alternative zur Zwei-Kannen-Methode zu schaffen. Obendrein wollte er verbessern, was ihn beim täglichen Gebrauch schon immer gestört hatte: Der heiße Inhalt sollte überall ohne Untersatz abgestellt werden können, ohne dabei hitzeempfindliche Unterlagen zu schädigen. Auf enge Papier-Teefilter oder Textilnetze, die er unansehnlich fand, wollte von Grolman bei der neuen Kanne verzichten. Obwohl sich seine Kunden längst an die Teeflecken auf seinen Zeichnungen gewöhnt hatten, durfte seine Teekanne im Gegensatz zu vielen anderen nicht tropfen. „Das hieß für mich sofort, der Teekörper sollte aus Glas sein“, erinnert sich der Designer. Zufällig entwarf er damals gerade Messestände, Musterhäuser und Innenarchitektur für den Laborglashersteller Schott Mainz, eine Nachfolgefirma jener Glaswarenfabrik, für die einst Wilhelm Wagenfeld sein mundgeblasenes Teeservice schuf. Obwohl auf diesem Gebiet noch unerfahren, erkannte von Grolman mit Gestalter-Blick, dass bei Laborgläsern das Problem des tropffreien Ausgießens weitaus besser gelöst war als im Haushalt. Kurzerhand übertrug er die präzisen Ausgüsse von Laborkolben auf das Teevergnügen. Schott half gern mit Prototypen aus besonders robustem, feuerfestem Duran-Glas. Tatsächlich ging die Rechnung auf. Die neue Glaskanne tropfte nicht.

TRIGGER: A LECTURE

Tassilo von Grolman hit upon his sensational teapot idea during a function at the Rosenthal Studiohaus in Frankfurt. That evening of autumn 1979 revolved around tea. Cultivation, processing, preparation, effect – tea experts examined the subject from every angle. Tassilo von Grolman, himself a passionate tea drinker, reviewed the design of teapots and was irritated to come to the conclusion that, still, really two teapots were needed for an optimum tea taste. Using this ancient method, tea is brewed in one teapot and then sieved through a strainer into a second heated teapot. Even classic teapots such as that of Walter Gropius or the aforementioned glass teapot of Wilhelm Wagenfeld were conceived in this way. The tiny tea strainers were simply intended to catch any remaining leaves from the first teapot. That same evening, with these reflections in mind, von Grolman decided to create an alternative to the two-teapot method.

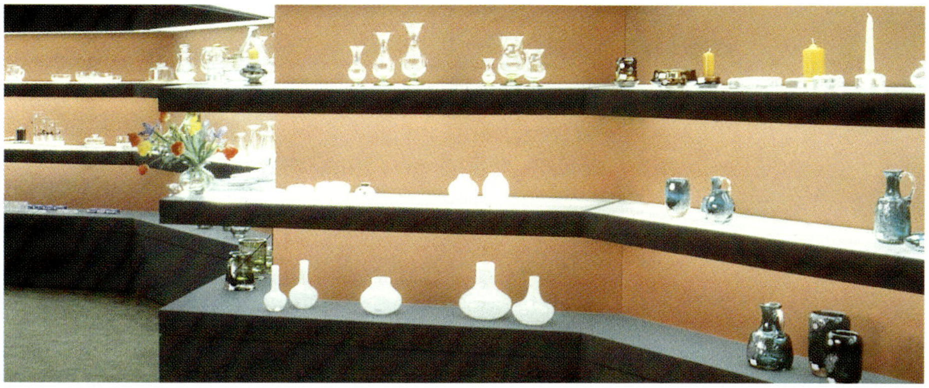

Tassilo von Grolman gestaltete in den siebziger Jahren Messestände für Schott. Regale, die sowohl von unten als auch von oben beleuchtet sind, rücken die Produkte ins rechte Licht.

In the seventies, Tassilo von Grolman designed exhibition stands for Schott. Shelves which are lighted from below as well as from above present the products in the best possible light.

Furthermore, he wanted to correct something that had always annoyed him in daily use: one should be able to put the hot contents down anywhere, without damaging heat sensitive surfaces. With the new teapot, von Grolman wanted to do away with thin paper filters or fabric netting, which he found unsightly. Although his clients had long been used to the brown stains on his drawings, his new teapot, in contrast to the many others, would not dribble. "For me that immediately meant that the body of the teapot should be made from glass" the designer remembers. By chance he was designing exhibition stands, show houses and room interiors for the laboratory glassware manufacturer Schott Mainz – a descendant of the glassware company for which Wilhelm Wagenfeld formerly provided his hand blown tea service. Although still inexperienced in this area, von Grolman's designer eye recognized that the problem of a drip-free pour could be better resolved using laboratory glass rather than household materials. In short, he transferred the precise spouts of laboratory retorts to the enjoyment of tea. Schott was happy to help with the prototype made of particularly robust and fire resistant Duran glass. It really did stack up. The new glass teapot did not drip.

Das allein war schon ein großer Gewinn gegenüber anderen Teekannenmodellen. Schlüsselgedanke des findigen Designers aber war: „Tee braucht Platz. Ohne Platz kann auch der beste Tee sein Aroma nicht entfalten." Von Grolman konzipierte also ein Edelstahl-Sieb, das nur unwesentlich kleiner als die Kanne selbst ist. Ein naheliegender Gedanke eigentlich, der dennoch in seiner Konsequenz ein Novum war. Herkömmliche Teeeier, -netze oder -filter sperrten das Teearoma allesamt durch ihre winzigen Dimensionen ein. Um das Sieb leicht zu füllen, mehr noch aber um die gequollenen Teeblätter später ohne Mühe wieder herauszubekommen, wählte von Grolman für Kanne und Sieb eine nach oben offene Halbkugelform. Diese war eine Reminiszenz an bauchige Kannenklassiker, durch ihre vollflächige Öffnung, dennoch ein formales wie ästhetisches Novum. Lange brütete von Grolman dann über dem noch immer offenen Problem der Hitzeregulierung zwischen Korpus und Stellfläche, denn er wollte seine Kanne überall abstellen können, „auch wenn sie 80 °C hat". Schließlich hob von Grolman den Glasbehälter einfach vom Boden weg, indem er ihn in ein Gestell aus Metallbändern einfasste. Zum Schluss brachte der Designer noch den Griff im Gestell unter und ein planer Deckel schloss die Halbkugelform seiner Teekanne nach oben hin ab. Etwa ein halbes Jahr nach jenem Teeabend im Rosenthal-Haus konnte von Grolman die erste Tasse Tee aus seinem Prototypen genießen.

This alone was already a major advantage over other models of teapot. The key thought of the resourceful designer was, however, "tea needs space. Even the best tea cannot develop its flavour without space." Von Grolman therefore created a stainless steel strainer, which was only marginally smaller than the actual teapot. This was actually an obvious idea but the consequences were nevertheless a novelty. The minute dimensions of traditional tea holders, nets or filters simply locked in the tea flavour. In order to make the strainer easy to fill or, rather, to enable easy removal of the swollen tea leaves after use, von Grolman chose a top opening hemispherical design for the teapot and strainer. This was reminiscent of classic bulbous teapots but its completely flat opening was both a technical and aesthetic novelty. Von Grolman pondered at length the still unresolved problem of heat regulation between the body of the teapot and the base, as he wanted to be able to put the teapot down anywhere "even when it is 80 °C." In the end, von Grolman raised the glass container off the ground by means of an encasing stand made from metal strips. Finally, the designer found space for the handle within the stand, and a plain lid at the top completed the hemispherical design of his teapot. Approximately half a year after that tea evening in the Rosenthal-Haus, von Grolman could enjoy the first cup of tea from his prototype.

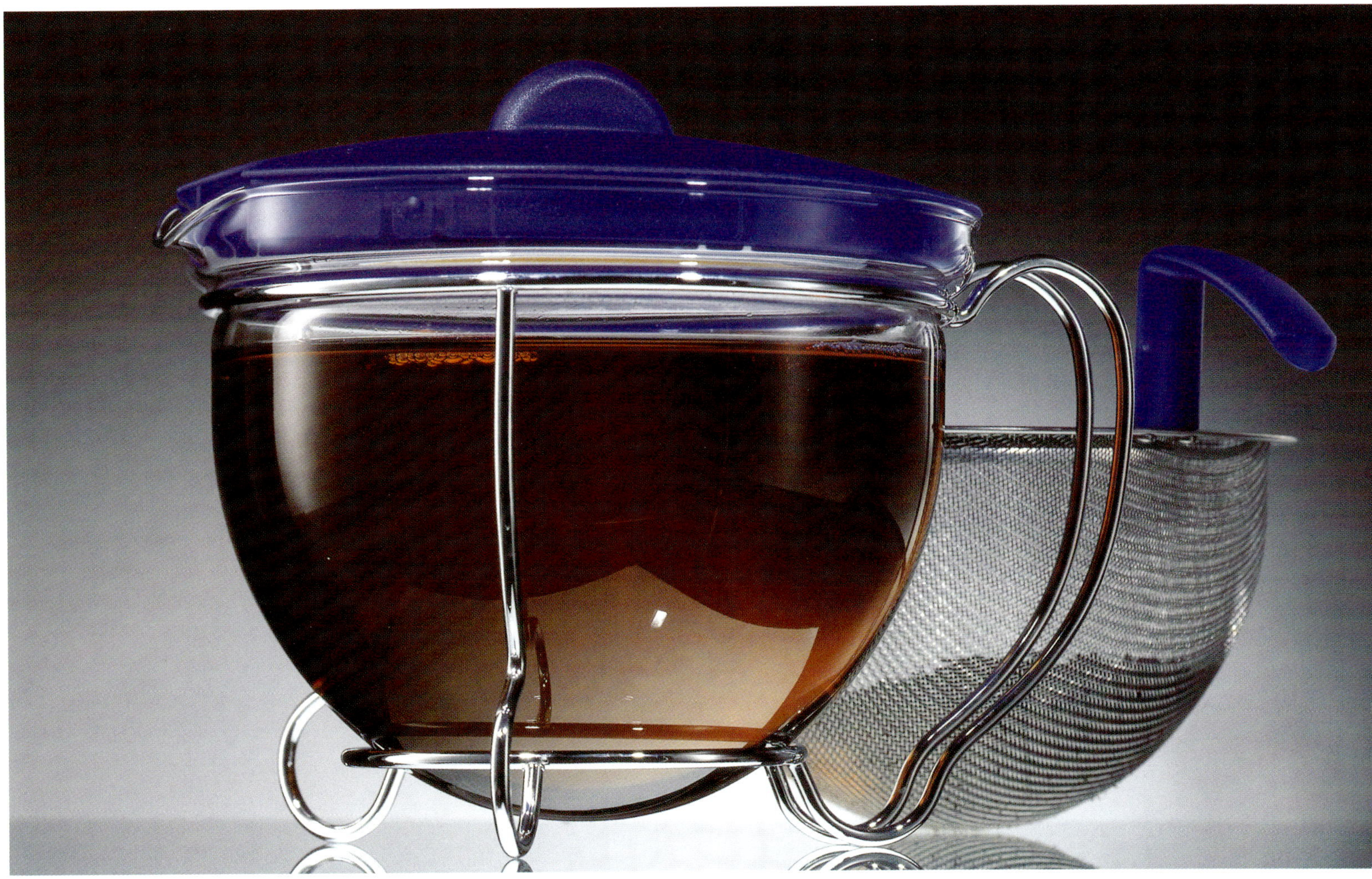

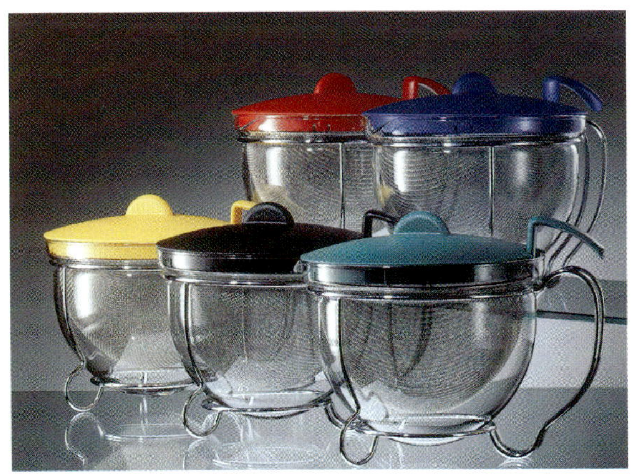

„Mono TeeColor" mit farbigem Kunststoffdeckel, 1993.

„Mono TeeColor" in den Farben Rot, Blau, Gelb, Schwarz, Grün.

"Mono TeeColor" with colored plastic lid, 1993.

"Mono TeeColor" in the colors red, blue, yellow, black, green.

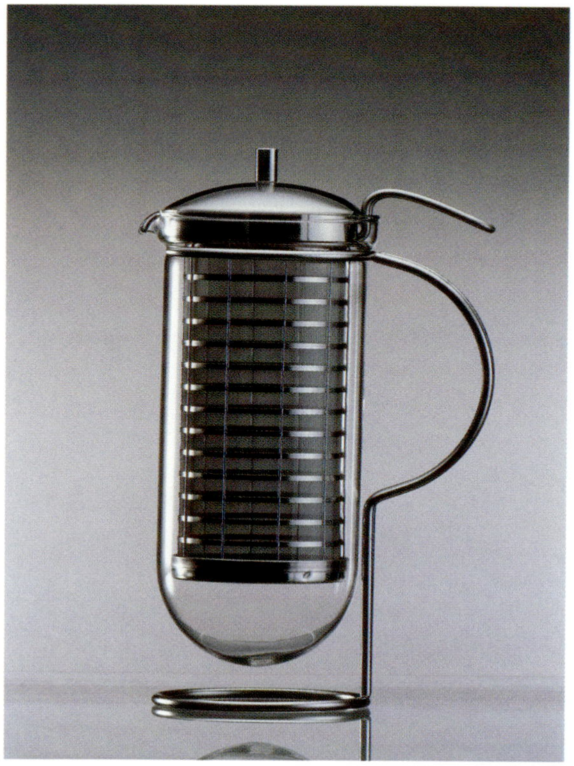
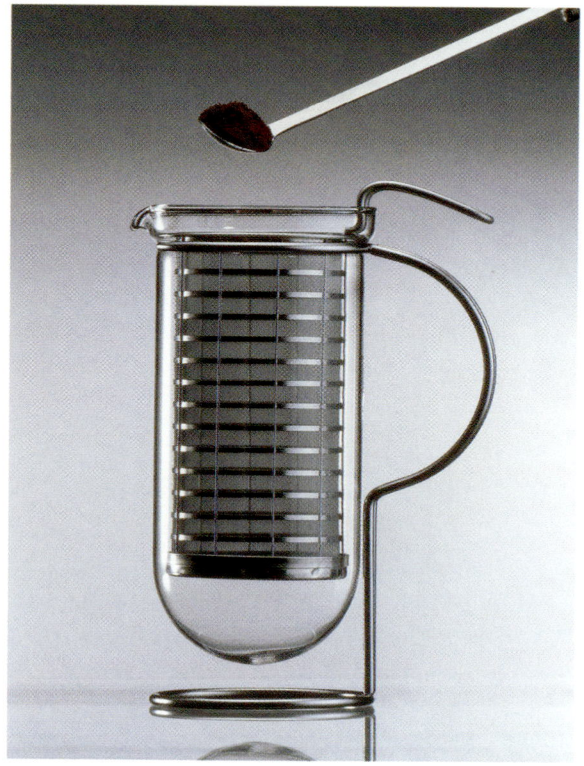
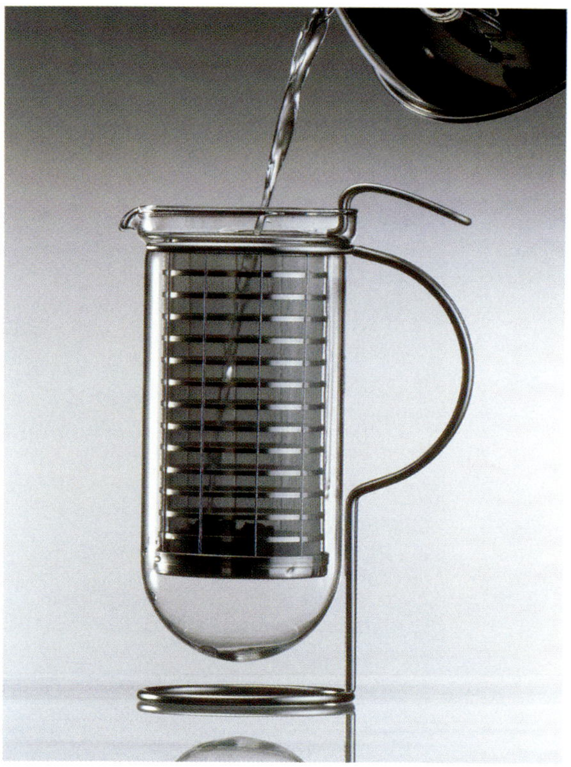
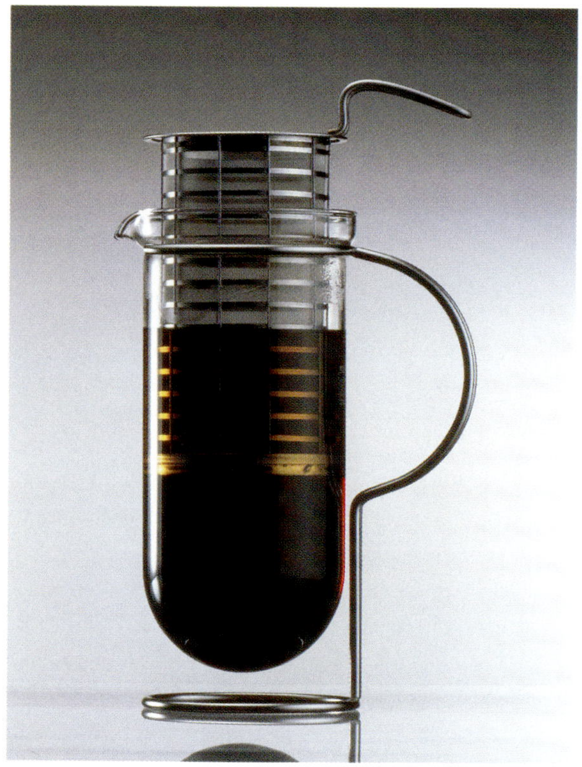

MARKTEINFÜHRUNG MIT HINDERNISSEN

So weit, so vielversprechend. Von Grolman war auf das Resultat seiner Überlegungen zu Recht stolz. Die Kanne erfüllte die anspruchsvollen Ziele des Designers perfekt. Ästhetisch brachte das Teegefäß High-Tech-Strenge und Yuppie-Coolness der achtziger Jahre auf den Punkt. Bei außergewöhnlichen Innovationen fragt man sich unwillkürlich, warum es so lange dauerte, bis jemand auf diese Lösung kam. Tassilo von Grolmans Kanne ist ein solcher Entwurf und zu Recht heute ein Klassiker. Kaum zu glauben ist deshalb, dass der Designer seinen Entwurf anbieten musste wie Sauerbier. WMF, Rosenthal, Melior, Melitta, Villeroy & Boch – „keiner wollte sie machen", resümiert von Grolman. Nach anderthalb Jahren fand sich endlich ein Metallwarenhersteller aus dem Bergischen Mettmann, der den Mut hatte, Neues zu wagen – Mono. Doch Mono konnte die Edelstahlbänder, die von Grolman in seiner Urform vorgesehen hatte, nicht fertigen. Erst Bodum produzierte die erste Version später als Modell „Chambord". Von Grolman überarbeitete also für Mono sein Ur-Modell, und im Frühjahr 1983 kam „Mono Tea 1" aus Edelstahldraht auf den Markt. Von bestechender Einfachheit, pur und leicht anzusehen, ist der Entwurf bis heute ein kommerzieller Erfolg. Dabei war die für eine Teekanne völlig neue technoide Anmutung zunächst für viele stark gewöhnungsbedürftig. Im Laufe der Jahre entstanden mehrere Varianten von „Mono Tea 1". Am erfolgreichsten ist bis heute „Filio" von 1987, mit der von Grolman eine weitere ästhetische Innovation gelang: Besonderer Clou ist der deutlich vom Boden abgehobene Glaskörper. Sein Verbindungselement zum Boden bildet gleichzeitig einen schwungvollen Henkel.

„Mono Cafino" als Metallvariante mit verschiedenen Siebeinsätzen für Mono Metallwarenfabrik GmbH, 1992.

"Mono Cafino" in metal variant with different strainer inserts for Mono Metallwarenfabrik GmbH, 1992.

MARKET ENTRY WITH OBSTACLES

So far, so promising. Von Grolman was rightly proud of the result of his deliberations. The teapot met the ambitious objectives of the designer perfectly. The tea receptacle aesthetically embodied the high-tech severity and yuppie coolness of the eighties. With such exceptional innovations one usually wonders why it took so long for somebody to find such a solution. Tassilo von Grolman' s teapot is one such design and today it is rightly considered a classic. It is hard to believe that the designer had to peddle his creation as if he were selling old beer. WMF, Rosenthal, Melior, Melitta, Villeroy & Boch – "none of them wanted to manufacture it," summarized von Grolman. Eighteen months later, a metal goods manufacturer from Bergisch Mettmann, who had the courage to dare something new, came forward – Mono. But Mono could not make the stainless steel bands that von Grolman had envisaged in his original design. Bodum produced the first version, later sold as the "Chambord" model. So von Grolman revised his original model for Mono, and in spring 1983 "Mono Tea 1" made with stainless steel wire, came onto the market. With captivating simplicity, pure and easy on the eye, the design remains a commercial success today. However, many people had to get used to the completely new techno appearance of a teapot. As the years went by, several variations on "Mono Tea 1" were introduced. The most successful to date is "Filio" from the year 1987, with which von Grolman succeeded in creating a further aesthetic innovation: a particular highlight is the glass body that is distinctly raised off the ground. The component with which it touches the ground also forms a curved handle.

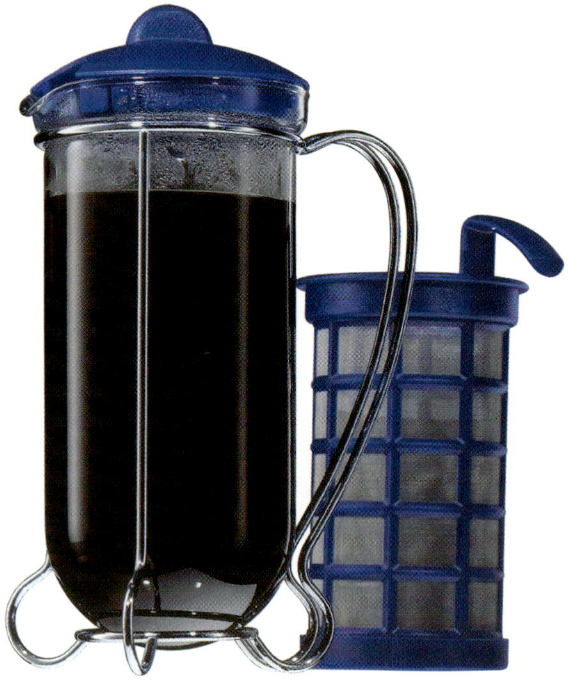

MAUERBLÜMCHEN CAFINO

Im Schatten der Teekannen geht eine weitere Innovation des Frankfurter Designers fast unter. 10 Jahre nach „Mono Tea 1" entstand die Kaffeekanne „Mono Cafino", der von Grolman nicht weniger als eine neue Art der Zubereitung mitgab. Laut eines Kaffeebreviers entsteht optimal schmeckender Kaffee, wenn man Kaffeepulver in eine Porzellankanne (!) gibt und siedendes Wasser darüber gießt, kurz umrührt, die Kanne mit dem Deckel verschließt und den Kaffeesatz zum Kannengrund sinken lässt – fertig. Allerdings gilt es, mit dem Kaffeesatz fertig zu werden, ohne dabei das Aroma zu beeinträchtigen, wie Kenner es zum Beispiel bei Kaffee ausmachen, der durch Filtertüten gelaufen ist. Man könnte nun meinen, nach Brühkaffee, Koch- und Espressokaffee, Goldfilter, Stempelkannen und Kaffeemaschinen wären die Zubereitungsmöglichkeiten von gerösteten Bohnen erschöpft. Weit gefehlt. Etwa 600 Jahre nachdem der Siegeszug des Kaffees von Äthiopien aus begann, übertrug Tassilo von Grolman 1992 sein bewährtes Teezubereitungsprinzip (Glaskörper plus nur knapp kleineres Sieb zum Einhängen) auf den Kaffee. „Cafino" mit der für Kaffeekannen typischen langgestreckt-zylindrischen Gestalt bekam ein nur knapp kleineres Sieb zum Einhängen. Nachdrücklich betont von Grolman den Unterschied zu Siebstempelkannen, denn für ihn ist das Stempelverfahren nicht ohne Risiko. Nach 100- oder 200-maliger Benutzung kann es nämlich passieren, dass beim Herunterdrücken des Stempels das Glas platzt, „entstanden durch Haarrisse, die beim Hinunterführen des Siebes an den zylindrischen Rand feinste Haarriefen im Glas erzeugen", führt er fachmännisch aus. Der Fall sei zwar selten, doch immerhin könne es außer zu Schnittverletzungen zu Verbrennungen ersten und zweiten Grades kommen. Im Gegensatz zu besagten Siebstempelkannen wird das eingehängte Sieb von „Cafino" nach einer kurzen Ziehzeit einfach herausgenommen. Dadurch bittert Tassilo von Grolmans Kaffee auch nicht nach. Die Vorteile von „Cafino" liegen auf der Hand, und endlich könnte das Gerät Charles de Talleyrands Ideal erfüllen, wonach Kaffee „heiß wie die Hölle, schwarz wie der Teufel, rein wie der Engel, süß wie die Liebe" sein muss. Leider kommt von Grolmans Erfindung für den französischen Botschafter, Außenminister und Lebemann rund 200 Jahre zu spät.

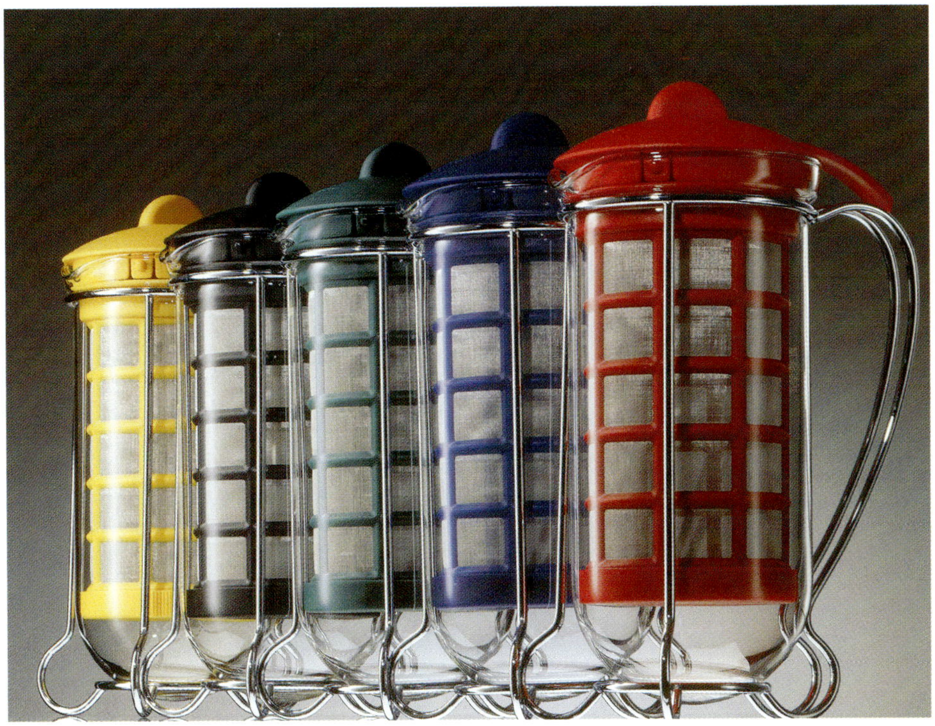

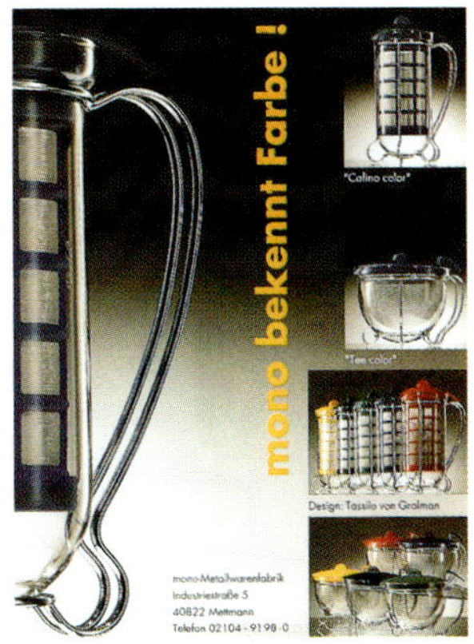

Linke Seite
„Mono Cafino Color" mit farbigem Kunststoffdeckel und -sieb, 1996.

Rechte Seite
„Mono Cafino" bekennt Farbe, 1996.

Left-hand side
"Mono Cafino Color" with coloured plastic lid and strainer, 1996.

Right-hand Side
"Mono Cafino" shows its true colours, 1996.

WALLFLOWER CAFINO

Hidden in the shadow of the teapots, a further innovation of the Frankfurt designer is almost lost. 10 years after "Mono Tea 1" the coffeepot "Mono Cafino" was created, which von Grolman gave no less a new method of drink preparation. According to one coffee guide, the optimum taste is obtained when coffee is added to a porcelain (!) coffee pot and boiling water is poured over it; it is then stirred for a short time, the lid is placed on the coffee pot and the coffee grounds are allowed to sink to the bottom – then it is ready. However, it is necessary to deal with the coffee grounds without spoiling the flavour, just as connoisseurs do with coffee that has gone through filters. It might be thought that after percolated coffee, boiled coffee, espresso coffee, gold filter, cafetières and coffee machines, the possibilities for preparing roast beans were exhausted. This is far from the truth. Around 600 years after the triumphal march of coffee from Ethiopia began, von Grolman transferred his principle for tea preparation (glass body with slightly smaller strainer hanging inside) to coffee. Typical for a coffeepot, "Cafino" had an elongated cylindrical form and a slightly smaller hanging strainer was added inside. Von Grolman emphasizes the difference to the cafetière strainer coffeepots, as he believes that the plunger mechanism of the cafetière is not without risk. After 100 or 200 occasions of use, the plunger could actually crack the glass when pushed down, due to "hairline cracks, which cause hairline grooves along the cylindrical edge when pressing down the strainer", he describes expertly. This would only happen rarely, but it could still lead to cuts and first or second degree burns. As opposed to the aforementioned plunger coffeepots, "Cafino's" suspended strainer is simply removed after a short percolating time. Also in this way, Tassilo von Grolman's coffee does not become bitter. The advantages of "Cafino" are clear and the appliance can finally attain Charles de Talleyrand's ideal, whereby coffee must be "hot as hell, black as the devil, pure as an angel, sweet as love." Unfortunately for the French ambassador, foreign minister and rake, von Grolman's invention is around 200 years too late.

NAIVITÄT UND GRÖSSE

YAKITORI

I Die Frühlingszwiebeln schälen und in 3–4 cm lange Stücke schräg abschneiden. Die Hühnerbrüstchen in 3–4 cm lange Stücke schneiden und in einen Topf geben. Sesamöl, Sojasauce und Sambal Oelek zufügen.

2 Die Knoblauchzehen schälen und mit dem Küchenmesser so lange hacken, bis sie eine klebrige Masse sind.

3 Mit dem Olivenöl und dem Salz würzen, gut unter das Hühnerfleisch rühren und mindestens 30 Minuten marinieren.

4 Die Frühlingszwiebeln und die Fleischstückchen fünf- bis sechsmal im Wechsel auf dünne lange Holzspieße stecken und so lange grillen, bis sie knusprig braun sind. Mit einem köstlichen, frischen Bier servieren.

3 Bund Frühlingszwiebeln

2 Hühnerbrüstchen

2 EL Sesamöl

60 ml Sojasauce

5 Knoblauchzehen

1 Msp. Sambal Oelek

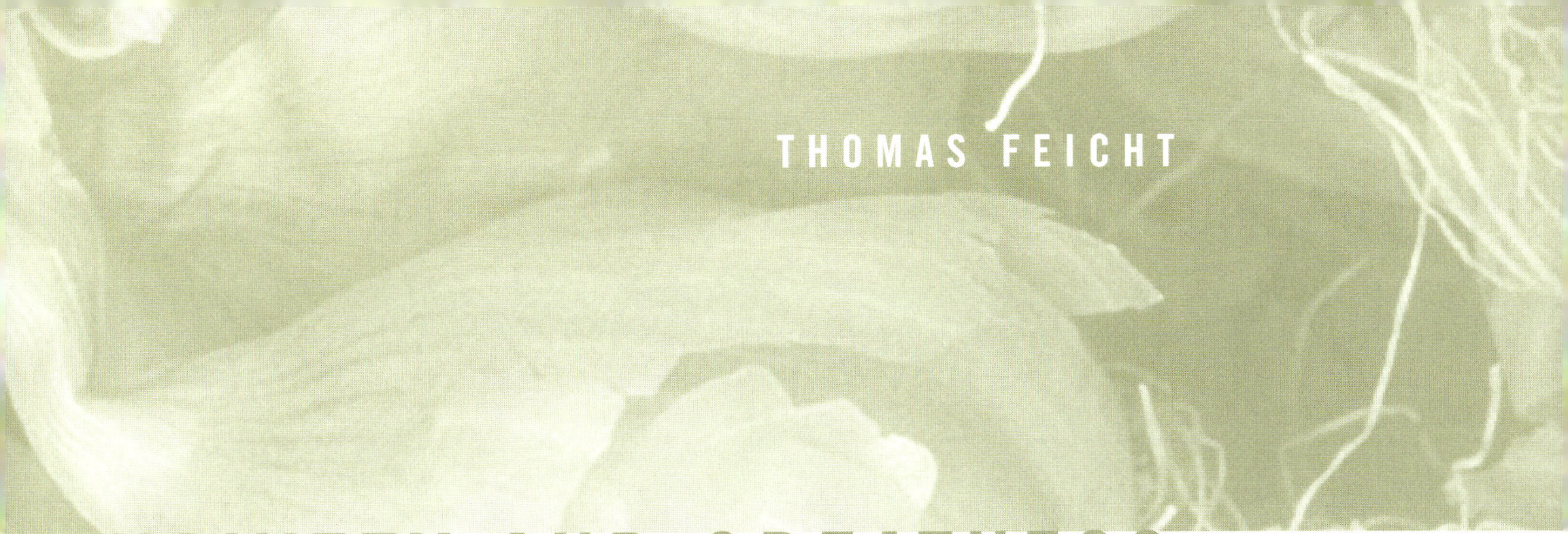

THOMAS FEICHT

NAIVETY AND GREATNESS

YAKITORI

1 *Peel the spring onions and cut at an angle into 3–4 cm long pieces. Cut the chicken breasts into 3–4 cm long pieces and put into a pan. Add sesame oil, soy sauce and sambal oelek.*

2 *Peel the cloves of garlic and chop with a kitchen knife until they turn into a sticky mixture.*

3 *Season with the salt and olive oil, mix well with the chicken and marinade for at least 30 minutes.*

4 *Push five to six pieces of spring onion and chicken pieces alternately onto wooden skewers and grill until they are crispy brown. Serve with a delicious cool beer.*

3 bunches spring onions

2 chicken breasts

2 tbsp sesame oil

60 ml soy sauce

5 cloves garlic

1 pinch sambal oelek

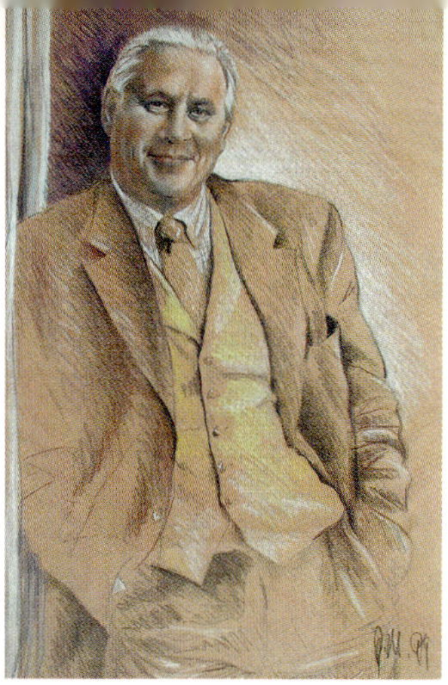

Linke Seite
Tassilo von Grolman, Portrait in Aquarellstifttechnik von Gabriele Maßmann, Mallorca, 1999.

Rechte Seite
Verpackungen für „Solano"-Glühbirnen (Woolworth), 1990.

Left-hand side
Tassilo von Grolman, portrait in coloured pencil technique by Gabriele Maßmann, Mallorca, 1999.

Right-hand side
Packaging for "Solano" light bulbs (Woolworth), 1990.

Entschuldigung. Aber ich muss etwas ausholen, einen ordentlichen Schlenker machen. Denn wie soll der geneigte Leser die Leistung, die Taten, das Werk, die Haltung eines Menschen verstehen, wenn wir ihn nicht leibhaftig machen. Und zwar in seiner Menschlichkeit, in seiner Persönlichkeit, in seiner Komplexität (so wäre das eigentlich eine Aufgabe für eine integrierte Kommunikation, weit über die Grenzen einer Monographie hinaus). Aber zurück zur Menschlichkeit, zum Menschen und Freund Tassilo, der fast naiv immer das Beste, das Schöne, das Gute, das Perfekte will. Ob er nun Gäste zusammenbringt (überlegt), ob und wie er sie bekocht (himmlisch), ob er die Tafel zelebriert (gekonnt), ob er den Wein wählt (stilsicher), ob er das Gespräch anregt (und irgendwann auf das kommt, was er eigentlich sagen und erreichen will). Es ist immer Überschwang, ja Freude dabei – vielleicht darf man sogar Liebe dazu sagen. Jetzt kennen Sie ihn ein klein wenig. Aber wir wollen ja von ihm lernen, in ihn schauen – oder sind Sie nicht neugierig? Also, all diese kindlichen Attribute hat er sich erhalten, sind sein Energiespender, lassen ihn die Welt immer wieder neu anpacken, neu sehen. Und das gepaart mit Erfahrung. Beruflich. Aber vor allem menschlich. Wobei wir bei dem zweiten Schlenker wären. Heute glauben wir an die Jugend (was ja gut ist), aber wir sollten nicht nur ihr glauben. Denn wir wissen ja, wie wir alle einmal angefangen haben: eingleisig, schmalspurig, egoistisch, vielleicht sogar engstirnig. Das war unser Recht. Denn wir waren Spezialisten. Mit den Jahren haben wir an Überblick gewonnen, sind menschlich gereift, können loslassen, zuhören und globaler denken, auch zum Nutzen anderer. Wir sind Generalisten geworden, ohne das spezifische Wissen, das Know-how vergessen zu haben. Ich denke, das ist Tassilo von Grolman: Ein Kind, das an das Gute glaubt, es sucht, mit menschlicher, erwachsener Reife, Größe und viel Wissen um das Leben, das Design und die Technik und was ich sonst noch vergessen habe. Aber langsam können wir ihn verstehen – hoffe ich doch – zum Beispiel seinen inneren Antrieb, den DDC (Deutscher Designer Club) zu gründen. Fast ganz allein gegen alle. Tassilo von Grolman ist von Hause aus Designer, hat seine Sporen in einer der besten Werber-Denker-Schmieden, der Agentur Lürzer, Conrad, verdient. Darüber – und über seine gute Arbeit – wurde er engagiertes Mitglied im ADC (Art-Directors Club Deutschland). Wie erkläre ich es kurz und bündig? Der ADC hat circa 290 Mitglieder. Davon zwei (!) Produkt-Designer. Tassilo war und ist einer davon und hat versucht, etwas zu ändern, wie zum Beispiel Design zum Thema des ADC zu machen sowie die Design-Kategorien in den jährlichen Wettbewerben ernster, wichtiger zu nehmen. Sogar ausbauen wollte er sie. Ausbauen nach dem Vorbild des angelsächsischen Art & Design Directors Club, wo es keine Trennung der Welten, keine Trennung von Werbung und Design gibt. Wäre dieses lobenswerte und völlig logische Unterfangen gelungen, dann gäbe es heute keinen DDC. Aber Kleingeister, In-Kategorien-Denker, Lobbyisten, Disziplin-Egoisten haben das verhindert.

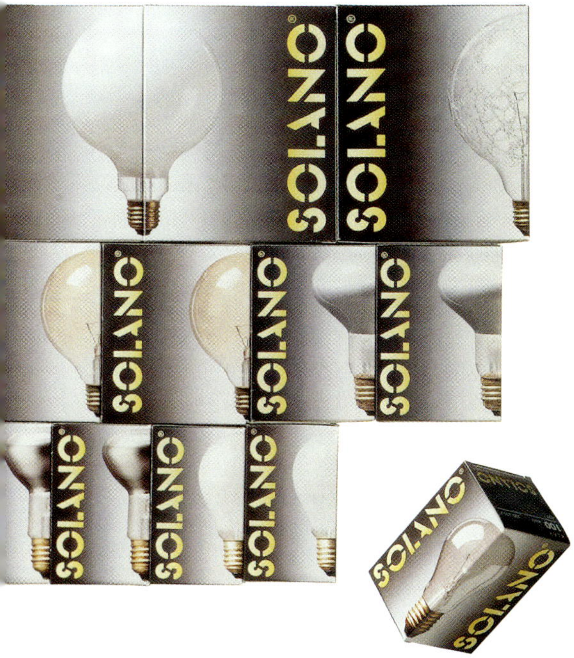

Sorry. But I have to go round the houses, to make quite a detour. How is the interested reader to understand the achievement, the deeds, the work, and the attitude of a person if we do not bring him to life? In all his humanity, his personality, his complexity (this would actually be an exercise in integrated communication, far exceeding the scope of a monograph). But back to humanity, to the man and friend Tassilo, who almost ingenuously wants the best, the beautiful, the good, the perfect. Whether he brings together (carefully selected) guests, whether and what he cooks for them (his cooking is heavenly), whether it is the celebratory laying of the table (in masterly fashion), or whether he is selecting the wine (with stylish assurance) or whether he is stimulating discussion (and, at some point, getting to what he really wants to say and achieve). Exuberance, even joy is always present – perhaps you could even say love. Now you know a little about him. But we want to learn from him, to see into him – and are you not curious? Well, he has maintained all these childlike attributes, they give him energy, allow him to continually grasp the world anew, see anew. And all this coupled with experience. Professional. But above all human. Which takes us to the second detour. Today we believe in youth (which is a good thing), but we should not only believe in it. We know how we all once started out: single-minded, limited in outlook, egotistical, perhaps even narrow-minded. That was our right. For we were specialists. Over the years, we have gained an overview, have matured as people, can let go, listen and think more globally, also for the benefit of others. We have become generalists, without forgetting the specific knowledge, the know-how. I think that is Tassilo von Grolman: a child, who believes in good and is searching for it with human and adult maturity, greatness and profound knowledge about life, design and technology and whatever else I may have forgotten. But slowly we can understand him – or so I hope – for example, his inner drive to found the DDC (German Designer Club). Almost completely alone against everyone else. Tassilo von Grolman is every inch a designer; he won his spurs in one of the best Advertiser-thinker forges, the agency Lürzer Conrad. Because of this – and his good work – he became a committed member of the ADC (Art Directors Club of Germany). How do I explain this in a nutshell? The ADC has around 290 members. Two (!) of these are product designers. Tassilo was, and remains, one of them and has tried to change things, for example by making design the theme of the ADC, by taking the design categories in the annual competitions more seriously, giving them greater importance. He even wanted to expand them. Expand them according to the example set by the Anglo-Saxon Art & Design Directors Club, where there is no division between the two worlds, no split between advertising and design. If this laudable and completely logical venture had succeeded, there would be no DDC today. But small-minded, one-track thinkers, lobbyists, discipline-egotists blocked it.

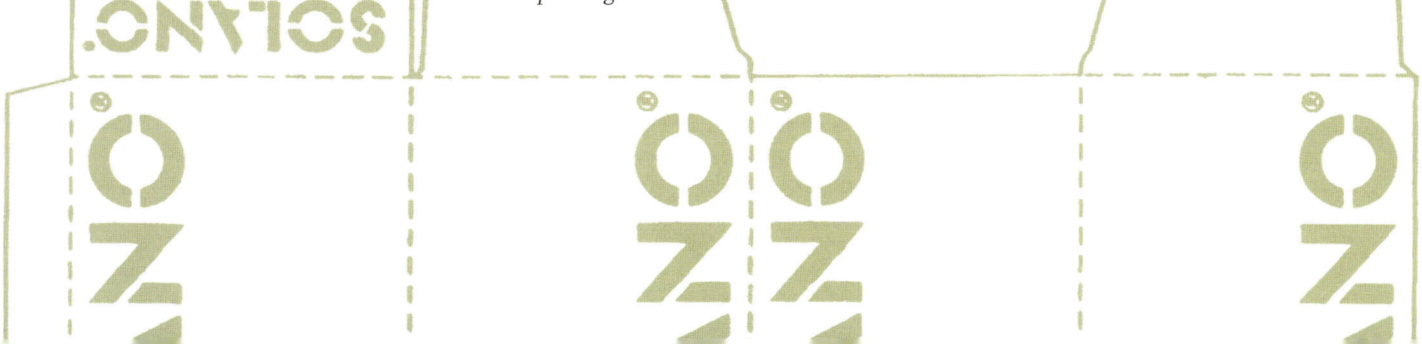

Tassilo von Grolman – wir kennen ihn ja inzwischen schon etwas genauer – hat ein paar Gleichgesinnte, sein Engagement und privates Geld genommen und 1989 den DDC gegründet. Dafür an dieser Stelle ein Extra-Lob! Es kann nicht groß genug sein. Denn es entstand ein Verein, eine Satzung, ein Büro, ein Corporate Design und zwei große Wettbewerbe (für „Designer bewerten Designer" und für „Vernetzte Kommunikation") mit anschließender Buch-Dokumentation – was für eine Energieleistung (von Geld gar nicht zu reden).
Aber manchmal – man nennt es das Glück des Tüchtigen – läuft die Welt auf einen zu. Und so war es hier. Die Welt wurde kleiner, komplexer, vernetzter. Der Begriff vernetzter, integrierter Kommunikation kam ins Rampenlicht sowie auch das Verständnis dafür, dass ein Unternehmen mit seiner Umwelt spricht, nicht nur über TV-Spots, nicht nur über Anzeigen und Prospekte. Sondern auch dadurch, was seine Produkte aussagen, welche Sprache Gebäude und Einrichtungen sprechen. Darüber hinaus rückte in den Mittelpunkt das Verständnis dafür, dass man all das und noch viel mehr steuern kann – ja muss.
Und da stand der DDC. Mit seinen Mitgliedern aus den Design-Disziplinen Architektur, Produkt-Design, Corporate Design, Werbung, Event, Promotion, aus den digitalen Medien und dem Messe-Design, um nur einige hervorzuheben. Folgerichtig haben wir gemeinsam – Tassilo als Treiber und guter Geist – den DDC im Jahr 1999 ganz scharf auf das Thema vernetzter Kommunikation und deren inhaltlicher und formaler Qualität positioniert, den ersten konsequenten Wettbewerb in diesem Bereich veranstaltet, Einsendungen eingefordert, die deutlich machen, wie viel mehr man für einen Kunden tun kann, wenn man (mit Partnern) konsequent über die Disziplin-Grenzen des Designers, des Architekten, des Werbers hinausdenkt und hinausrealisiert.
Im Jahr 2000 geht dieser Prozess dynamisch weiter: neues Erscheinungsbild, neuer Vorstand, Frankfurt wieder als Nukleus, Internet, Kooperation mit dem verlag form. Eine wichtige Kooperation für den DDC. Denn die Ziele des DDC lassen sich nur dann erreichen, wenn sie in einer qualifizierten Öffentlichkeit von Fachleuten und Meinungsführern maximal transportiert werden. Der verlag form bietet den Interessen des DDC eine optimale Plattform, öffnet für öffentlichkeitswirksame Buchprojekte (zum Beispiel eine Ausgabe mit den Gewinnern der DDC-Wettbewerbe) den Weg in den Buchhandel, stellt in seinen Titeln Möglichkeiten für die gezielte Öffentlichkeitsarbeit des DDC in bestehenden und neuen Medien bereit. Also: Auch hier ist ein weiterer bedeutender Baustein auf dem Weg zur total vernetzten Kommunikation für den DDC selbst, für seine Mitglieder, seine Anliegen, seine Haltung vorhanden. Und wieder zeigt es sich, wie wichtig eine Lebensgeschichte und nicht nur das Können ist. Tassilo als Produkt- und Corporate Designer und als Werber konnte diese Strömung erfühlen, erkennen, kanalisieren und mitschwimmen. Persönlich denke ich, dass dies eine große Bestätigung und eine kleine Sternstunde für seinen Einsatz ist, also rundherum ein Erfolg.
Aber das Leben bietet ja nicht nur Siege. Eine handfeste Niederlage für Tassilo von Grolman ist sein bisher erfolgloser Kampf für ein Designer-Urhebergesetz. Wie man an diesem Beispiel sieht, kämpfen Designer nicht nur um gute Gestaltung, sondern auch um die eigentliche Anerkennung des Kerns ihrer Arbeit: des Urheberrechts.

Linke Seite
DDC Hauptjurierung 1992. Der Präsident Tassilo von Grolman erläutert Jurymitgliedern die Funktionsweise des Exponates.

Rechte Seite
DDC-Buch Nr. 1, 1993.
DDC-Buch Nr. 2, 1996.

Left-hand side
DDC main judging 1992. The president Tassilo von Grolman explains the function of the exhibit to members of the jury.

Right-hand side
DDC-book No. 1, 1993.
DDC-book No. 2, 1996.

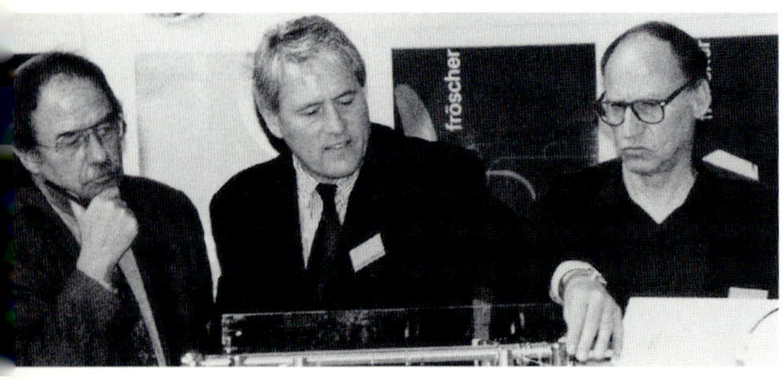

Tassilo von Grolman – we know him rather better now – took a few like-minded individuals, his commitment and his private money and founded the DDC in 1989. Special praise is due at this point! This praise cannot be high enough. Thus a club, a constitution, an office, a corporate design and two major competitions ("Designers judge designers" and "Networked Communication") with the resultant bookwork were established – what an energetic (not to say financial) performance. But sometimes – you could say fortune favours the brave – the world comes to you. And this was the case. The world became smaller, more complex, more networked. The term 'networked, integrated communication' emerged into the limelight, together with the understanding that a company must talk to its environment, not just through TV ads, not just through advertisements and brochures. But also through the statements that its products make and that which is communicated through buildings, fixtures and fittings. Beyond that, the understanding that all of this, and much more can – and must – be steered, became a focal point.
And there stood the DDC. With its members from the design disciplines of architecture, product design, corporate design, advertising, event management, promotion, from digital media and exhibition design, just to mention a few. Consequently in 1999 we – with Tassilo as driving force and good spirit – positioned the DDC strongly with the theme of networked communication, its quality of content and form. We held the first competition in this area, called in submissions, which show how much more can be done for a client when you (with partners) consistently think and deliver beyond the bordered disciplines of the designer, the architect, the advertiser.
In the year 2000, this process continues dynamically: new image, new board, Frankfurt again the nucleus, Internet, co-operation with the publisher "verlag form". An important co-operation for the DDC. Because the aims of the DDC can only be achieved if they are communicated at best within a qualified audience of experts and opinion leaders. The publisher "verlag form" offers an ideal platform for the interests of the DDC. It opens the door to the book trade with book projects that are particularly effective in terms of public relations (such as an edition of winners of the DDC competitions) and through its publications makes available possibilities in existing and new media for the targeted public relations work of the DDC. This is a further significant building block on the road to totally networked communication for the DDC itself, for its members, for its concerns, for its position. And again it goes to show how important the story of one's life is and not only ability. Tassilo, as a product and corporate designer and as an advertising man, could feel and recognize the flow, channel it and swim with its tide. Personally I think that this is a great confirmation and a small period of greatness for his endeavours, an all-round success.
But life does not just offer victories, and one substantial defeat for Tassilo von Grolman has been his struggle for a designer copyright law, which remains unsuccessful to date. As this example shows, designers not only fight for good design but also for the actual recognition of the core of their work: copyright.

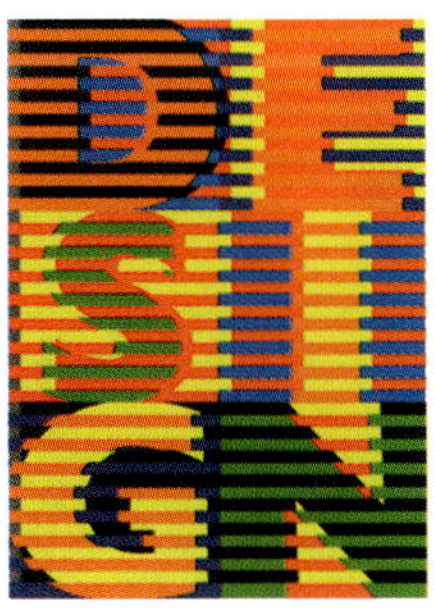

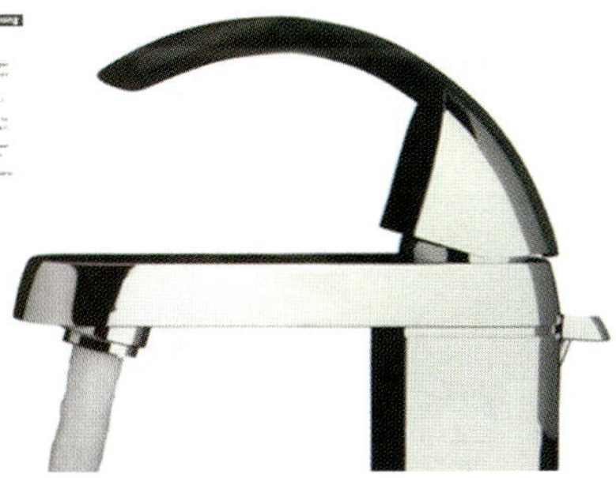

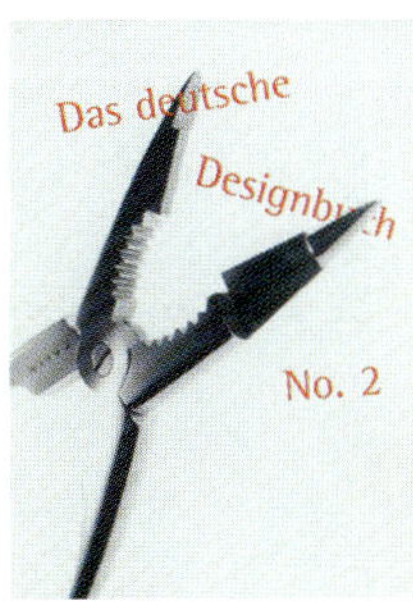

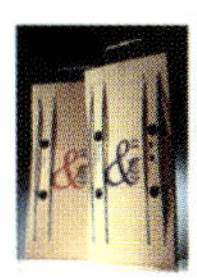

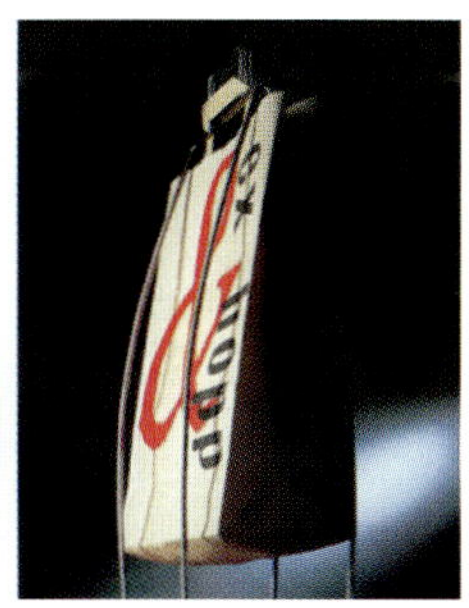

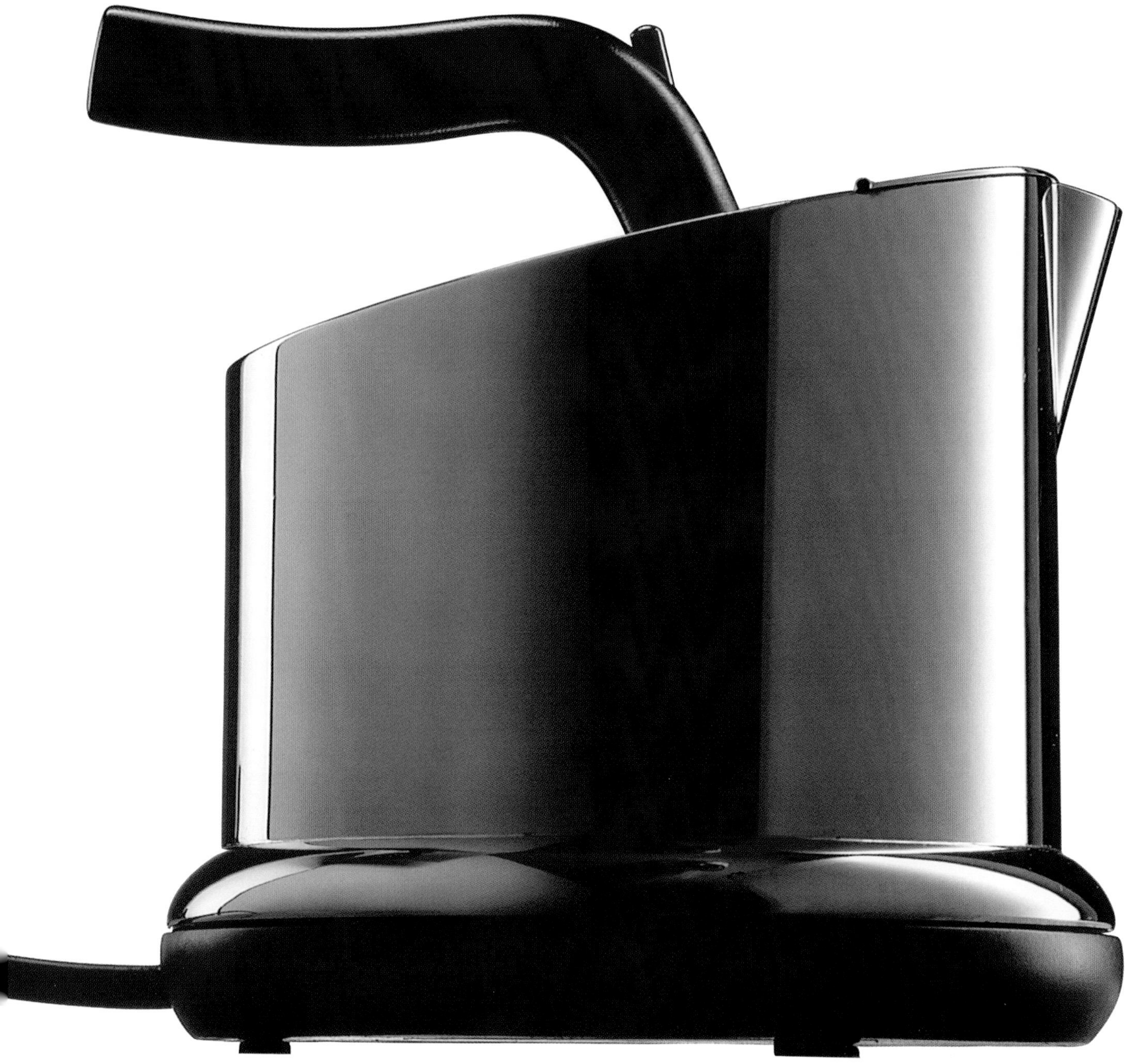

Elektrischer Wasserkocher, Studie für WMF, 1988.

Electric water boiler, study for WMF, 1988.

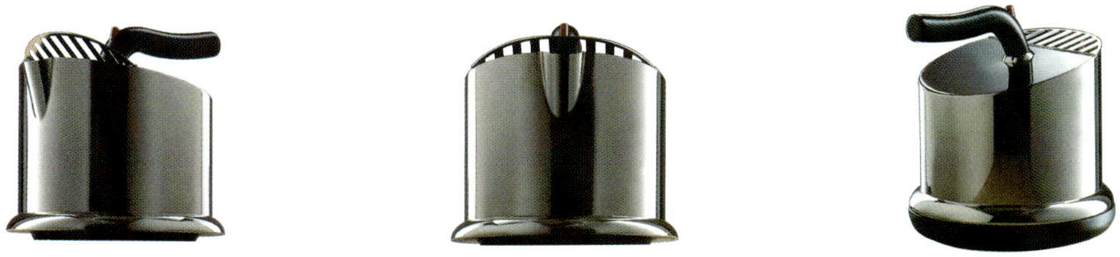

Musiker haben es geschafft – und zwar gemeinsam mit der Industrie, die von ihnen lebt – ordentliche Tantiemen, sprich Copyrights oder Nutzungsrechte zu bekommen. Verdienen dürfen doch alle. Oder nicht? Warum nicht gleiches Recht für alle, für alle Designer, die ursächlich unser aller Leben gestalten. Und da möchte ich noch einmal unsere vernetzte, integrierte Welt strapazieren. Musiker haben es also angeblich geschafft. Architekten haben eine Gebührenordnung, aber warum gibt es Ideenwettbewerbe mit bis zu 400 oder 600 Teilnehmern – wer erkennt da noch die Idee oder was ist der Schweiß dem Auftraggeber wert? Für Künstler, Illustratoren, Fotografen, Schriftsteller ist das Recht auch schon ausgefochten. Produkt-Designer werden da unterschiedlichst abgefunden, eine einheitliche Regelung gibt es nicht. Schon gar nicht, wenn es um Plagiate geht (da werden nur Richter und nicht Designer erhört). Werbung. Ja, da wird leider oft nur die Oberfläche gesehen und der schnelle Gag oder gar nur die Umsetzung bezahlt. Aber die Wertsteigerung einer Idee wird in der Regel nicht honoriert. Also brauchen wir Betroffene (im doppelten Sinne des Wortes) und Kämpfer wie Tassilo von Grolman, die Verbände, Unternehmer, Kollegen, die Öffentlichkeit wachrütteln. Es macht doch keinen Sinn, uns wie Zitronen auszuquetschen. Und vielleicht kommen wir ja doch einmal zu einer Gleichbehandlung der Idee, zu dem Wert einer Idee über die einzelnen Berufsbilder hinweg. Somit gibt es keinen Grund sich auszuruhen, müde zu werden.

Musicians have managed to gain proper royalties – and done so together with the industry that feeds them – for example copyright or usufruct. Everybody is allowed to earn. Are they not? Why not have the same law for everyone, for all designers, who causally design our whole lives. And here again I would like to task our networked, integrated world. Well, musicians have supposedly done it. Architects have a fee structure, yet why are there competitions for ideas with up to 400 or 600 participants – who then recognizes the idea and what is the hard work worth to the client? Artists, illustrators, photographers, authors have also fought their battle for that right. Product designers have been fobbed off in the most diverse ways; there is no single rule. Particularly not when the question concerns plagiarism (where only judges and not designers are heard). Advertising. Yes, unfortunately, here often only the surface is seen, and the quick gag or even only the implementation is paid for. But the value added to an idea is generally not financially recognized. Therefore we need people who are affected by this and fighters such as Tassilo von Grolman, to shake the associations, businessmen, colleagues and the public awake. It makes absolutely no sense to squeeze us like lemons. And perhaps we will arrive at equal treatment of the idea after all, at the value of an idea over and above the individual job description. So there is no reason for us to rest, or tire.

Womit wir wieder den Bogen geschlagen hätten hin zu der Persönlichkeit, dem Menschen Tassilo von Grolman. Er ist Er-Fühler, Macher, Umsetzer, Antreiber, Medium, Denker, Katalysator, Opfer, Kämpfer, Vorbild, Unzufriedener, Ruheloser, Schön- und Poltergeist, Frohnatur und Sensibelchen, ein guter Freund und konsequenter Feind. Er ist ein Findling in unserer Zeit, in der Jetzt-Zeit, die nach Haltung sucht, um sie ringt, wo die Menschheit Unterhaltung, scheinbar nur noch Unterhaltung sucht. Aber da möchte ich gerne wieder zur Musik eine Parallele ziehen und Fritz Rau (Vordenker und Galionsfigur der Konzertveranstalter) zitieren: „Unterhaltung ist per se nichts Schlechtes, wenn sie nicht unter der Haltung des Künstlers ist.“ Welch ein schöner Gedanke, welch treffendes Wortspiel. Die Vielfalt, die Individualität ist erlaubt – aber jeder einzelne sollte dazu stehen können und (wieder Fritz Rau) nicht Lieschen Müller oder den Ruhrkumpel zitieren müssen, um seine Arbeit zu rechtfertigen. Das sollten wir uns täglich fragen – und Tassilo von Grolman handelt sogar, ohne sich das fragen zu müssen, denn er ist eben ein Urgestein und handelt aus sich, aus seiner Lebens-Haltung heraus. Er will Unter-Haltung auf hohem Niveau, so nahe wie möglich am Ideal. Und wenn wir im Hier und Jetzt leben, das Leben, unser Leben gestalten und es wertvoll machen – ja dann – dann sollten wir uns eine Scheibe von Tassilo von Grolman abschneiden. Guten Appetit.

With that we have once again made the connection to Tassilo von Grolman, the personality and human being. He is a feeler, a maker, an implementer, a driver, a medium, a thinker, a catalyst, a victim, a fighter, a role model, not satisfied, restless, an aesthete, a poltergeist, cheerful and sensitive, a good friend and tough enemy. He is a foundling of our times, in the here and now, that is looking for dignity, struggling for it, where humankind is seeking entertainment and seemingly only entertainment. But here I would like to draw another music parallel and quote Fritz Rau (forward thinker and figurehead of concert promoters): "There is nothing wrong with entertainment per se, if it is not beneath the dignity of the artist." What a lovely thought. Variety, individuality 'is allowed – but every individual should hold to it and (again according to Fritz Rau) not have to quote the man on the street or Joe Bloggs to justify his work. We should ask ourselves that question daily – and Tassilo von Grolman even operates without having to ask himself the question, because he is such a rock and acts according to his creed for life. He wants entertainment of the highest level, as close as possible to the ideal. And as we live in the here and now, shaping life – our life – and making it worthwhile, yes then, then we should take a leaf out of the book of Tassilo von Grolman. Enjoy your meal.

Etiketten für „Bad Pyrmonter“ Mineralwasser, 1989.

Labels for "Bad Pyrmonter" mineral water, 1989.

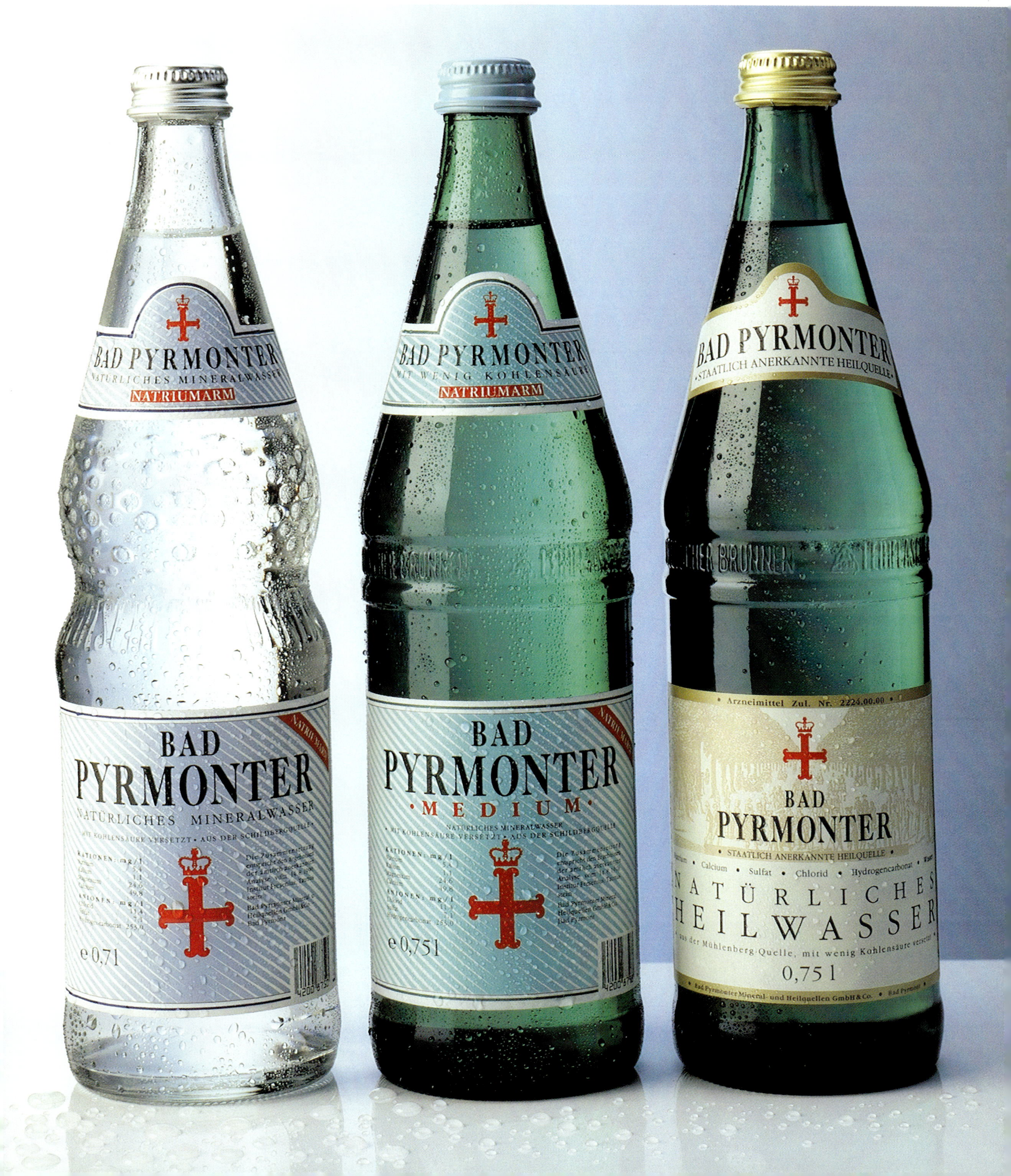

BAD PYRMONTER
NATÜRLICHES MINERALWASSER
NATRIUMARM
BAD PYRMONTER
NATÜRLICHES MINERALWASSER
e 0,7 l
BAD PYRMONTER
MIT WENIG KOHLENSÄURE
NATRIUMARM
BAD PYRMONTER
MEDIUM
e 0,75 l
BAD PYRMONTER
STAATLICH ANERKANNTE HEILQUELLE
Arzneimittel Zul. Nr. 2224.00.00
BAD PYRMONTER
STAATLICH ANERKANNTE HEILQUELLE
Calcium • Sulfat • Chlorid • Hydrogencarbonat
NATÜRLICHES HEILWASSER
0,75 l

GESTALT NEU DENKEN

Petersilie

Sauerampfer

Schnittlauch

Basilikum

Rauke/Rucola

frischer Koriander

Kapern

2 Zehen Knoblauch

4–5 EL Essig

1 EL Senf

1/4 l Olivenöl (Extra virgine)

Salz

Pfeffer

Kümmel

4 Eier

1 kg Kartoffeln

GRÜNE SOSSE À LA TASSILO

1 Für die Soße: Kräuter unter fließendem Wasser gründlich waschen und fein hacken. Ich persönlich gebe die Kräuter dafür in eine Küchenmaschine. Knoblauch schälen und zusammen mit Essig, Senf und Olivenöl zugeben.

2 Alles in der Küchenmaschine oder in einer Schüssel verrühren und mit Salz und Pfeffer abschmecken.

3 Inzwischen die Eier hart kochen. Kartoffeln mit der Schale und dem Kümmel in reichlich Salzwasser weich kochen.

4 Die hart gekochten Eier schälen, halbieren und auf den Tellern anrichten. Grüne Soße à la Tassilo darüber geben und servieren.

5 Grüne Soße à la Tassilo kann natürlich auch zu Tafelspitz, gekochtem Schinken, gekochter Kalbszunge, Hausmacher-Sülze, geräuchertem Lachs oder geräucherter Forelle gereicht werden.

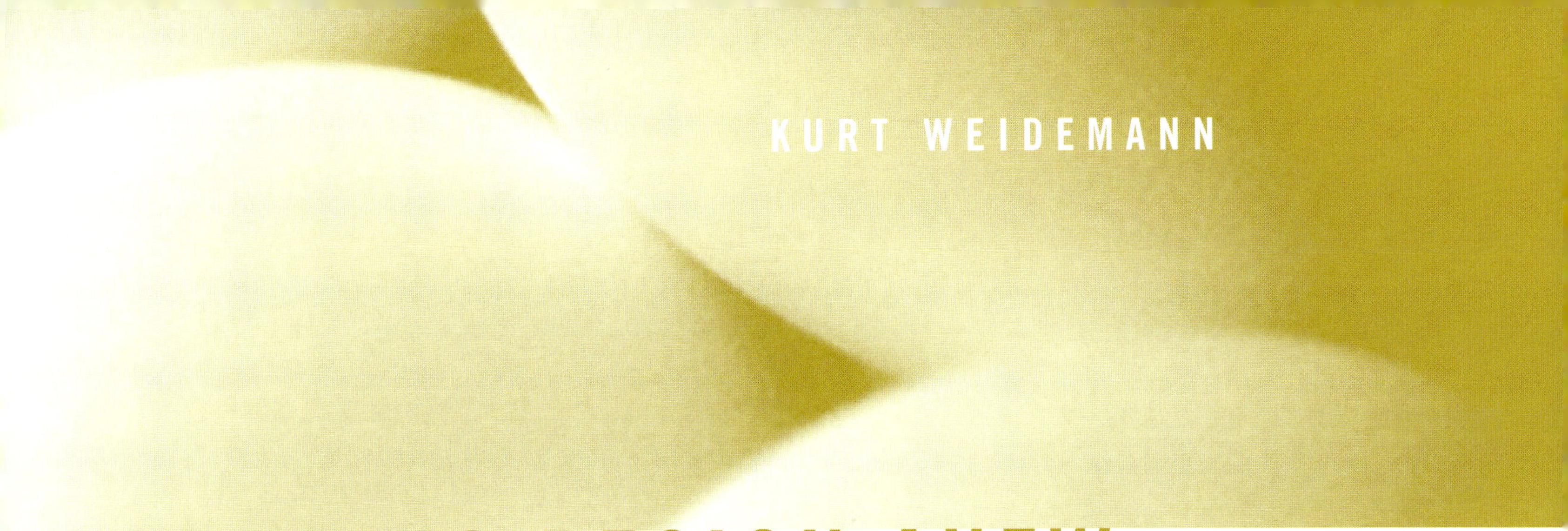

KURT WEIDEMANN

THINKING DESIGN ANEW

Parsley

Sorrel

Chives

Basil

Rucola

Fresh coriander

Capers

2 cloves garlic

4–5 tbsp vinegar

1 tbsp mustard

1/4 l olive oil (extra virgin)

Salt

Pepper

Caraway seeds

4 eggs

1 kg potatoes

GREEN SAUCE À LA TASSILO

1 *For the sauce: wash the herbs thoroughly under running water and chop finely. I personally put the herbs into a food processor to do this. Peel the garlic and add together with the vinegar, mustard and olive oil.*

2 *Mix everything in the mixer or in a bowl and season with salt and pepper.*

3 *Meanwhile hard-boil the eggs. Boil the potatoes in their skins together with the caraway seeds in plenty of salted water.*

4 *Peel the hard-boiled eggs, cut them in half and arrange on plates. Pour over the green sauce à la Tassilo and serve.*

5 *Green sauce à la Tassilo can of course also be served with boiled beef, boiled ham, boiled tongue, home made brawn, smoked salmon or smoked trout.*

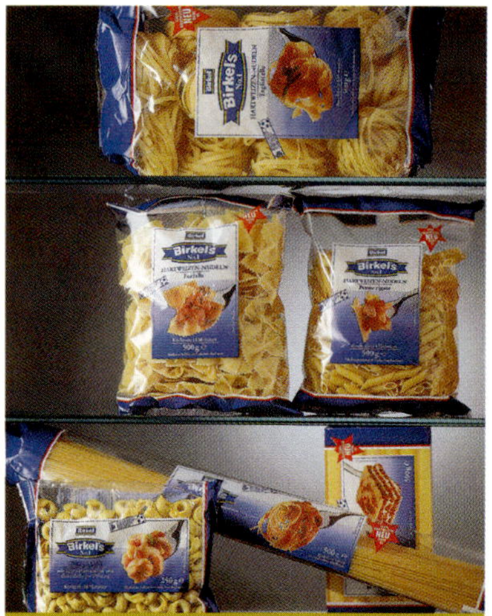

Nach der Entschlüsselung des Erbgutes möchte der Mensch nun Gott spielen. Unzufrieden mit dem Ergebnis von sieben Schöpfungstagen – nach wie vor von Schnupfen und Mücken und Migräne geplagt – will er jetzt einen achten Reparaturtag nachschieben. Beim Ausverkauf der Schöpfung soll das verbesserte Angebot Schwung und neue Hoffnung bringen. „Na, na", sagt da der Bauer und bleibt skeptisch. Und der Pfarrer warnt vor Risiken und Nebenwirkungen und nimmt seinen lieben Gott in Schutz, der den Menschen nicht nur als genetischen Rohstoff gedacht hat. Bevor wir dann zu einem genetisch veredelten Produkt kommen, versuchen wir uns – so gut es geht – im Diesseits einzurichten. Und wir müssen uns das Bewusstsein wach erhalten, dass es in unseren Breitengraden vergleichsweise sehr komfortabel zugeht. Es gibt auch noch einen „Rest der Welt". Wenn auch die Verheißungen eines in allen Ersatzteilen auswechselbaren, kaum noch störanfälligen Menschen wie Glocken klingen, schließt das doch das Nachdenken über Lebenswert und Lebenssinn nicht aus. Und den Menschen muss man weiterhin essen und schlafen lassen, dem wenig verhohlenen Drang sich zu vermehren nachgeben, ihn wärmen oder kühlen, behausen und versichern. Darüber hinaus ist er veranlagt, zu versuchen, all diese Lebensumstände zu erleichtern, zu verbessern, zu vereinfachen. In der Verwertungskette „menschliches Dasein" müssen alle Glieder gleich stark miteinander verbunden sein. Das ist so einsichtig wie der Satz: Es bleibt alles so, wie es niemals war.

After the decoding of genes, man wants to play God. Dissatisfied with the results of seven days of creation – still tormented by the same sniffles and midges and migraines – he now wants to add an eighth day for repairs. When selling off creation, the improved offer should bring zest and new hope. "Well, well", says the farmer and remains sceptical. And the cleric warns of risks and side effects and defends his dear God, who did not think of man only as genetic raw material. Before we arrive at a genetically improved product, we try – as far as possible – to equip ourselves in this life. And we have to keep reminding ourselves that we have a comparatively cushy life in this neck of the woods. There is also still a "rest of the world." Even if the promise of a human whose spare parts are replaceable and who is scarcely susceptible to faults sounds wonderful, this does not exclude reflections on the value and the point of life. And man must always be able to eat and sleep, to give in to the barely suppressed urge to breed, to be kept warm or cool and to have a roof over his head and to be insured. Above all he has a leaning to try to alleviate all the circumstances of life, to relieve, to improve, to simplify. In the food chain of "human existence", all links must be equally strongly connected. That is just as understandable as the sentence "Everything remains as it never was".

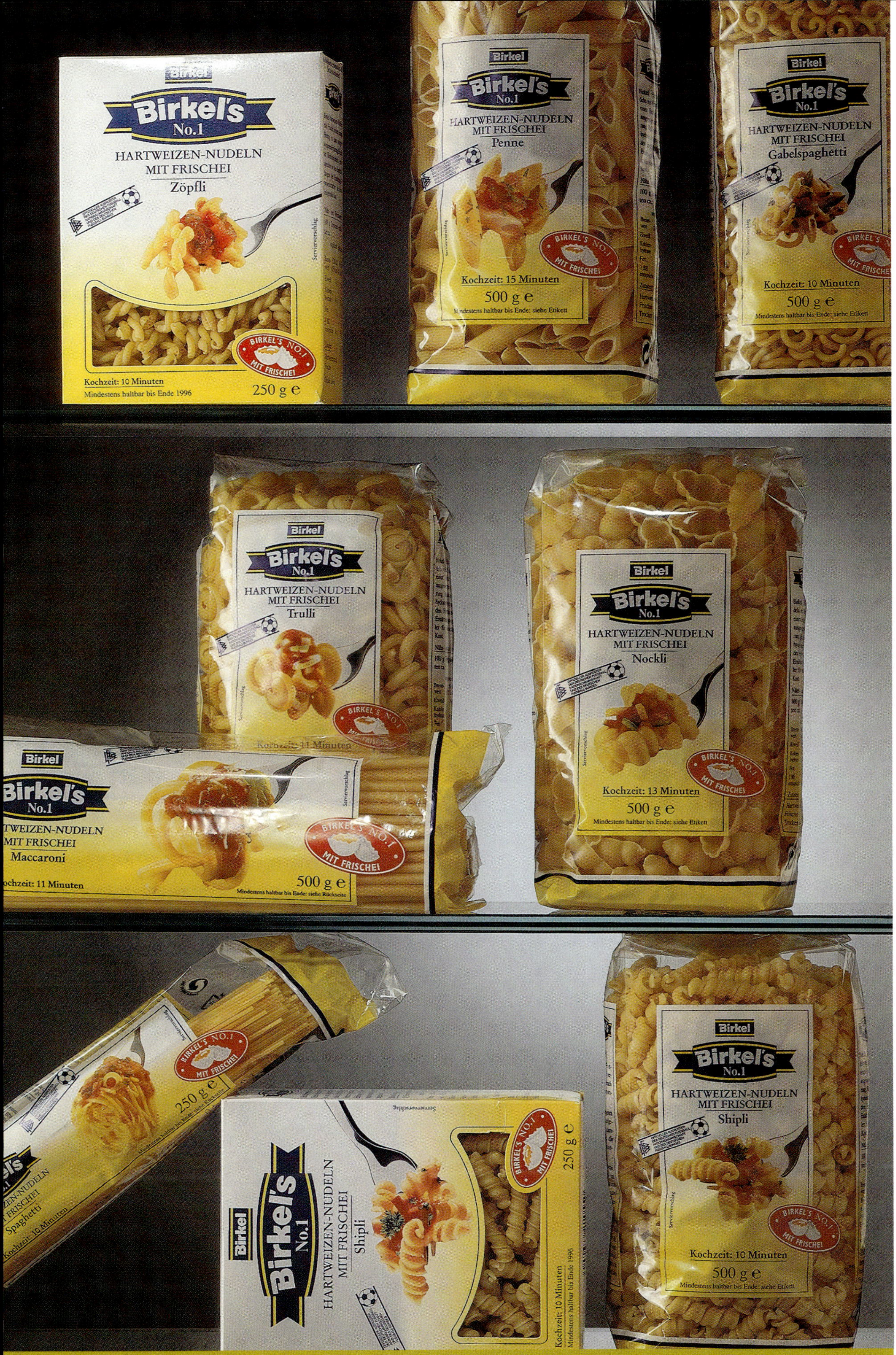

Verpackungen für Birkel-Nudeln, 1993.

Packaging for Birkel noodles, 1993.

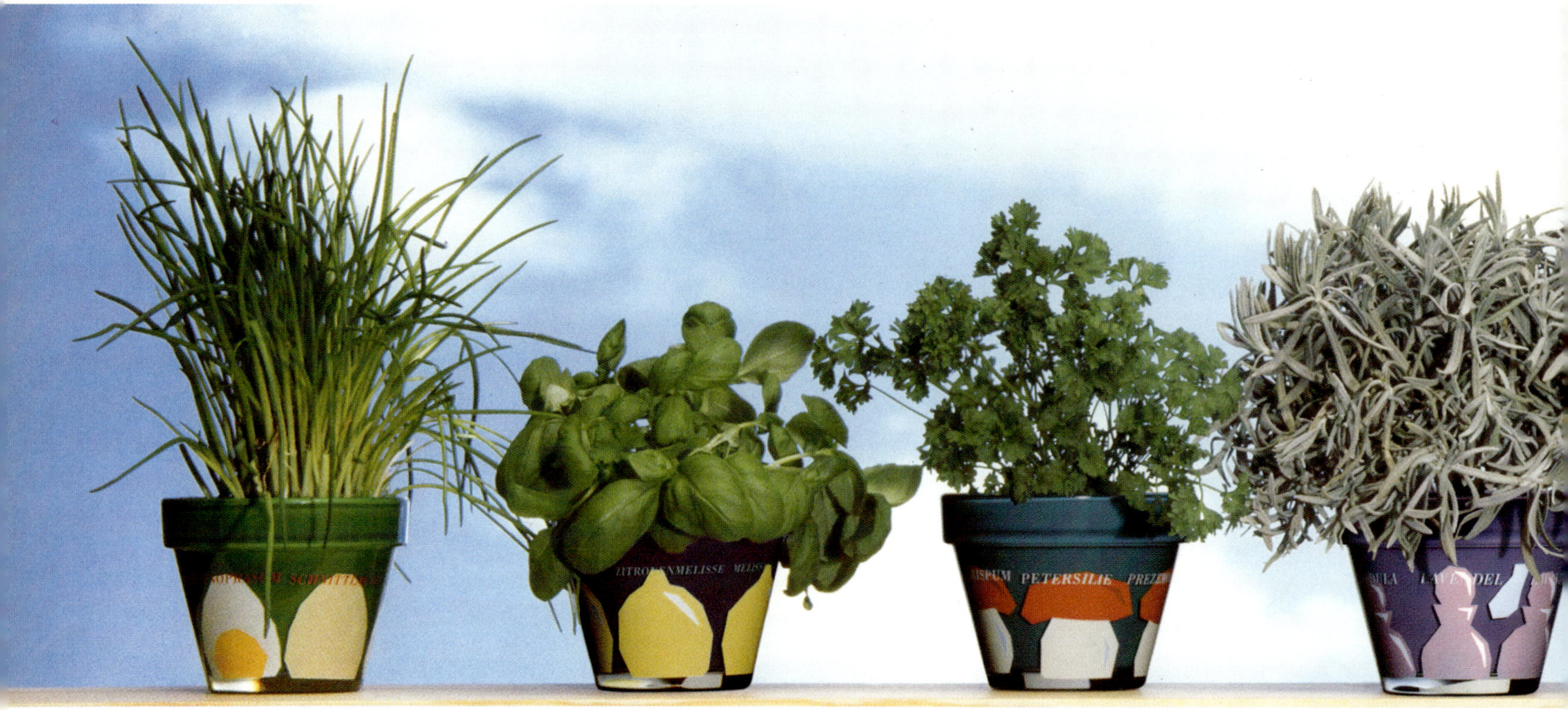

Oben
Dekore für Ritzenhoff „Herbs-Collection", 1997.

Rechte Seite
Give-away „Frankfurter Grüne Soße" für Commerzbank/Hausinvest, 1992.

Above
Decors for Ritzenhoff "Herbs-Collection", 1997.

Right-hand side
Free gift "Frankfurter Grüne Soße" Commerzbank/Hausinvest, 1992.

In einer längst mobilisierten und schon weit mediatisierten Welt, in der nichts mehr klar, einfach und unerheblich ist, ist das Verständigen nicht leichter geworden. In einer Welt, in der sich Leitbilder und Wertvorstellungen ständig verschieben, ist es schwer, einen Standort zu finden, ihn zu beziehen und zu vertreten. All das entbindet uns aber nicht von der Notwendigkeit, vor-, mit- und nachzudenken. „Denken heißt verändern." Und „verändern" heißt heute: neu gestalten, nicht kosmetisch umschminken, sondern eine Gestalt neu denken. Hinterher rennen, kopieren oder schlicht verschlafen, das reicht nicht. Dafür muss man einen Wert haben – einen Zusammenhalt finden – und einen Nutzen stiften. Von Berufs wegen kümmern sich darum die Designer verschiedener Sparten. Eine bildhauerische Kunstform in unabänderliche technologische Gegebenheiten zu gießen, war noch der Bildungsauftrag der Designosaurier. Ihre Spitzenprodukte haben Eingang in die Museen gefunden. Heute kann man auf dem Gestaltungscomputer alle Möglichkeiten und Unmöglichkeiten einer Form durchspielen. Die Hand rutscht mit zusammengewachsenen Fingern über der Maus hin und her und lässt die Variantenschleuder mal machen. Streichholzkopfgroße Chips können bereits Funktionen übernehmen, die im mechanischen Zeitalter noch handgreifliche Abmessungen hatten. Eigentlich könnten der Phantasie wieder große Wirkungsmöglichkeiten eröffnet werden. Aber eine verkopfte Herstellerwelt mit zahlengewohnten Entscheidungsträgern setzt Grenzen.

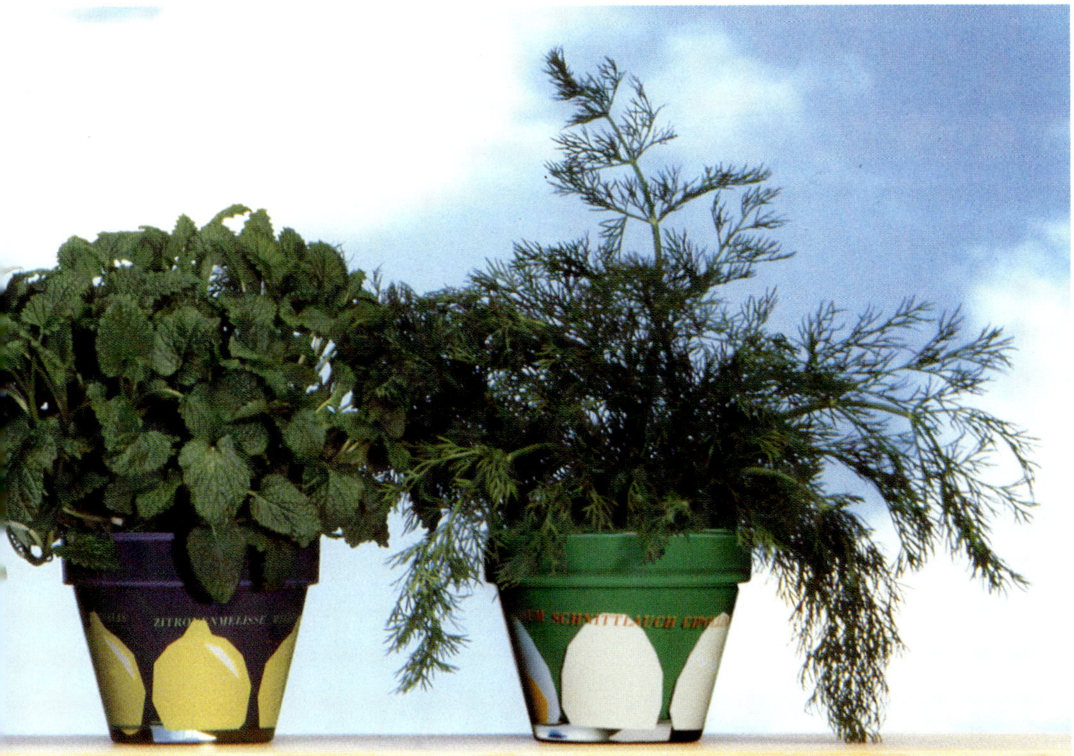

In a long mobilized and, already, broadly mediatised world, in which nothing is clear, simple and unimportant any more, communication has not become easier. In a world in which role models and moral concepts are constantly shifting, it is difficult to find an orientation to adopt and stand for. All of this does not, however, release us from the necessity to think ahead, to follow a train of thought or to reflect. "To think means to change." And "change" today means: to design from scratch, not to make cosmetic changes but to think up something new. It is not enough to follow, copy or simply sleep through something. One must have a value – find cohesion – and endow a use. On account of their profession, designers of diverse disciplines concern themselves with this. The training brief of the designosaurs was still to pour a sculptural piece of art into an unalterable technological mould. Their top products have found their way into museums. Today all the possibilities and impossibilities of a shape can be played out on the design computer. The hand, with fused-together fingers, slides to and fro over the mouse as the variations are spewed out. Chips the size of matchstick heads can already perform functions that in the mechanical era were still manual. This could really again open up great fields of activity for the imagination. But a hardheaded manufacturers' world – with decision-makers used to numbers – defines the limits.

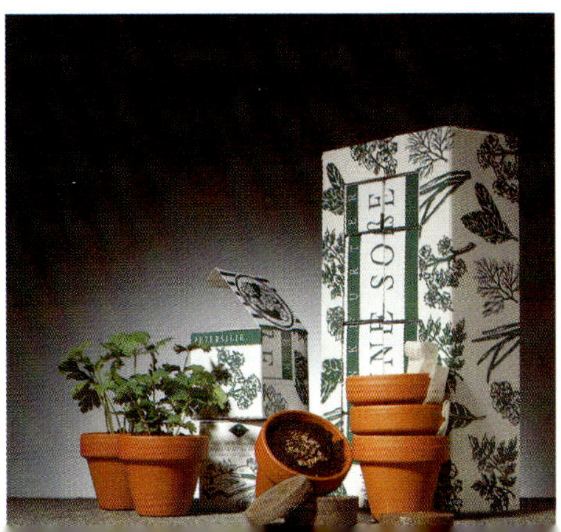

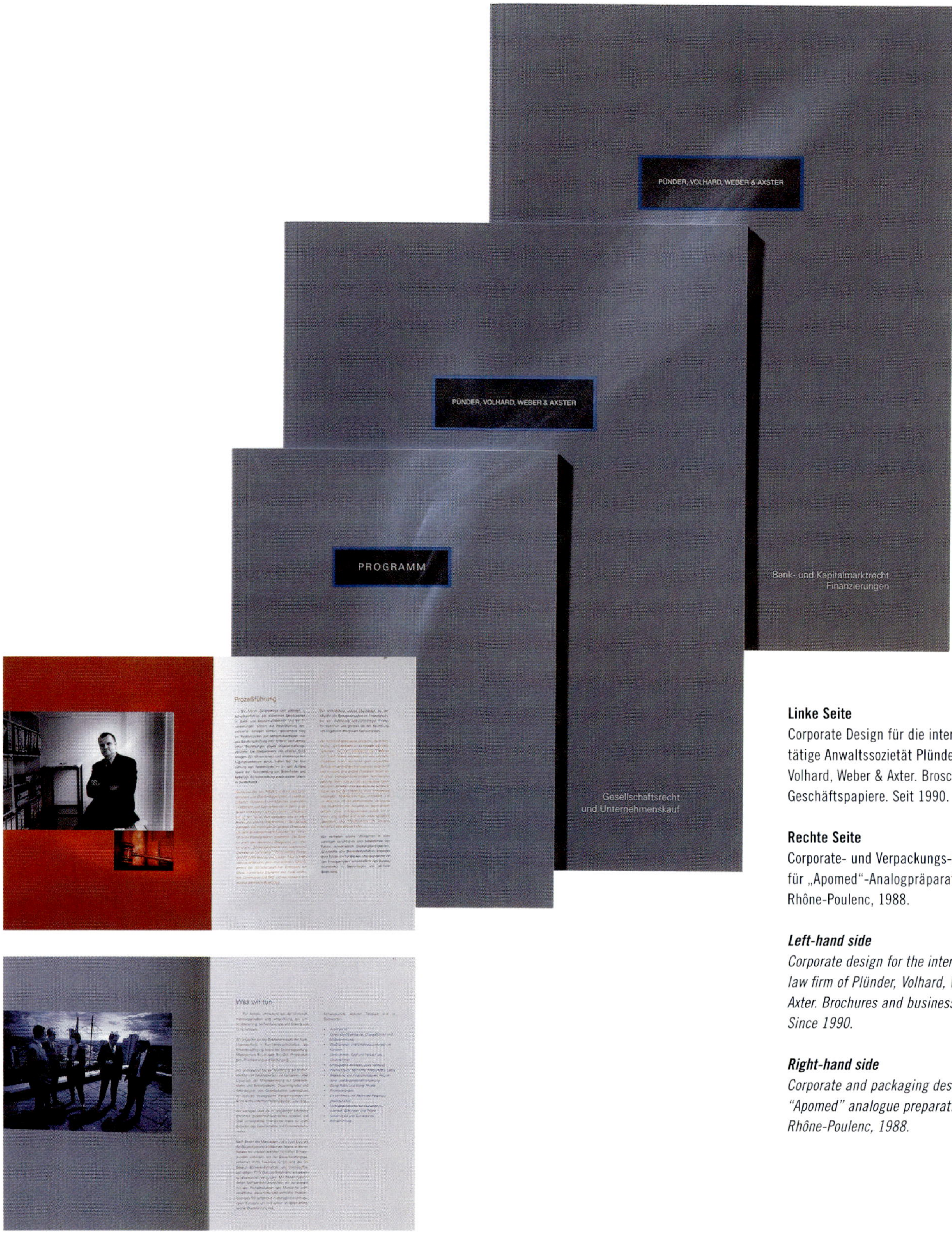

Linke Seite
Corporate Design für die international tätige Anwaltssozietät Plünder, Volhard, Weber & Axter. Broschüren und Geschäftspapiere. Seit 1990.

Rechte Seite
Corporate- und Verpackungs-Design für „Apomed"-Analogpräparate von Rhône-Poulenc, 1988.

Left-hand side
Corporate design for the international law firm of Plünder, Volhard, Weber & Axter. Brochures and business papers. Since 1990.

Right-hand side
Corporate and packaging design for "Apomed" analogue preparations by Rhône-Poulenc, 1988.

Nun kommt noch der Spaß mit dem Cyberspace dazu und baut uns eine virtuelle Welt ohne Bodenhaftung. Mit gelegentlich abartigen Folgen und Verwirrungen. Da wird die Frage nach Typen, die mit beiden Beinen auf Mutter Erde stehen und einen Dienst leisten, verständlich. Und da ist auch Tassilo von Grolman zu finden. Der Versuch, Designberufe in der Kunst anzusiedeln – was ganz selten möglich ist – hat etwas Vortäuschendes. In Kunst und Design ist gleichermaßen Kreativität gefragt, nur: die des Künstlers ist frei vagabundierend, die des Designers ist eng zweckgerichtet und hat durch einsichtiges, verbesserndes, erfinderisches, hochfunktionales Gestalten einer Verkaufsidee zu dienen. Mit Tassilo von Grolman kommt zu dieser Richtigkeit auch noch Charme und Flair, kommt ein „Ankommer" hinzu. Er ist detailverliebt, überlässt nichts dem Zufall, kann kalkulieren und selektieren, setzt Prioritäten, kennt die Zielgruppen, macht nicht was er will, sondern will, was er macht. Das sind Eigenschaften, die selten auf einem Stamm wachsen und sich oft diametral gegenüber stehen. Aber hier versteht einer, die strenge Logik des Kopfhirns mit der krausen Logik des Bauchhirns („Fuzzy Logic") in Harmonie zu bringen. Nicht zufällig ist er ein Meisterkoch, denn auch dort ist Erfindergabe gefragt. Entdeckungen gehören der Menschheit. Erfindungen gehören dem Erfinder. Im Zeitalter der elektronischen Betriebssysteme und der beliebigen Manipulierbarkeiten geht das Urheberrecht in die Binsen. Man sollte genügend Geld mitbringen, um seine Erfahrungen bezahlen zu können. Fortschritt heißt: Feuer weitergeben, nicht Asche bestatten. Über unser Behagen oder Unbehagen geben wir uns erst Rechenschaft, wenn sie im Extrem auffällig werden. Das beschäftigt von Grolman, das Erleben führt an das Denken, das Andersdenken, das Neudenken und endet bei Tea-Caddies, porzellanenen Dosen, Krümelmonstern, die nicht nur das Brot schützen, sondern auch das Schneidemesser bereit halten und die Brösel auffangen. In allem begegnet man der Lebenshaltung eines kultivierten Understatements, das Gestalt gewinnt.

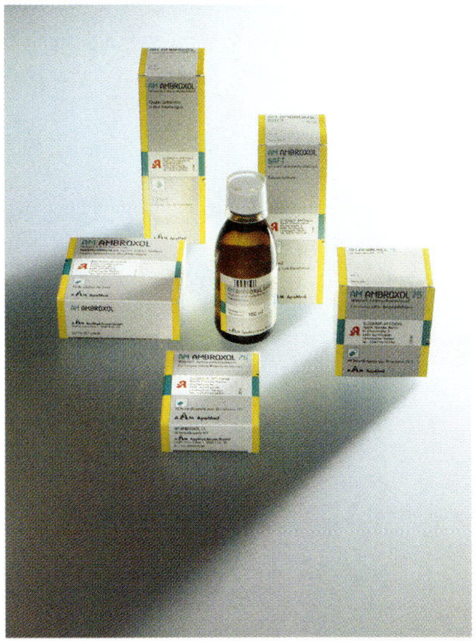

Now along comes the fun of cyberspace and builds us a virtual world without gravity. With occasional abnormal consequences and confusions. So the request for a type of person who has both feet firmly on the ground and provides a service becomes understandable. And here too, Tassilo von Grolman can be found. The attempt to fit design professions into art – which is seldom possible – is somewhat misrepresentative. A similar degree of creativity is demanded in both art and design, but: that of the artist can rove freely whereas that of the designer is tightly bound to a purpose and has to realise a sales idea through an understandable, improving, innovative and highly functional design. With Tassilo von Grolman charm and flair also add to this precision, an "arriver" is added. He displays a love of detail, leaves nothing to chance, can calculate and select, sets priorities, knows the target groups, does not do what he wants but wants what he does. These are properties that are rarely found together in one body, and are often diametrically opposed. But here somebody understands how to harmonize the strict logic of the head-brain and the muddled logic of the stomach-brain ("fuzzy logic"). It is not by chance that he is a master cook, as this also demands the gift of invention. Discoveries belong to mankind. Inventions belong to the inventor. In the age of electronic operating systems and arbitrary manipulability, copyright is given the boot. One should bring enough money to pay for one's experiences. Progress means: pass on the fire, do not bury the ashes. We only take notice of our contentment or discontent when it becomes extremely conspicuous. This is what concerns von Grolman, experience leads to thought, thinking differently, new thinking, and ends with tea caddies, porcelain jars and crumb monsters, which not only protect the bread but also store the knife ready for use and catch the crumbs. In all these things, we are confronted with the lifestyle of a cultivated understatement – design wins.

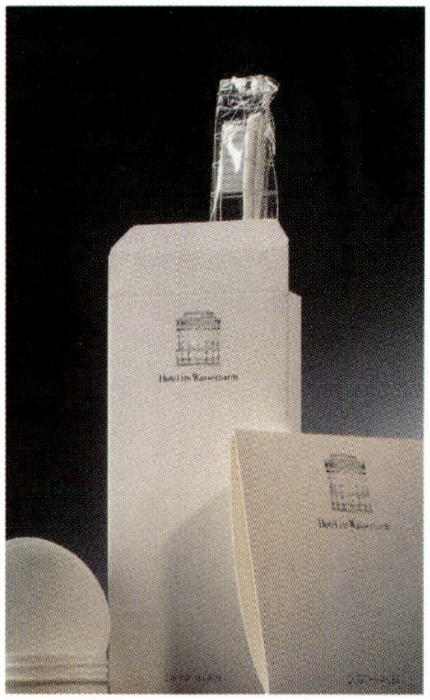

Linke Seite
Verpackungen für das von André Putman ausgestattete „Hotel im Wasserturm" Köln, 1990.

Rechte Seite
Porzellan-Teedose für „Twinings-Collection 2006", eine limitierte Künstleredition: Alessandro Mendini, Alan Fletcher.

Left-hand side
Packaging for "Hotel im Wasserturm" Cologne, furnished by André Putman, 1990.

Right-hand side
Porcelain tea caddy for "Twinings-Collection 2006", a limited artists' edition: Alessandro Mendini, Alan Fletcher.

Als Produzent bleibt sein Studio nicht beim Produkt stehen, sondern macht auch die Packung, die Verkaufshilfen, die Werbemittel. Als Agenturmann hat er Marketing-Erfahrung. Im freien Beruf, als Chefdesigner und als Mitbegründer und Vorstand des kleinen, aber feinen Deutschen Designer Club (DDC, 1989) hat er eine allseitig abgesicherte Berufspraxis, die auch noch durch aktive Tätigkeit nicht nur als Produktdesigner, sondern auch als Grafik-, Verpackungs-, Corporate-, Ausstellungs- und Architektur-Designer festgezurrt wird. Damit ist der Horizont dieser Berufssparten abgeschritten. Die Seefahrt hat seinen Rundblick geöffnet. Der Designerberuf hat sich aus der Enge des Wohlgestalters, Schönermachers und Verkaufsförderers befreit in die Dimensionen einer Sozialverantwortung. Das geht auch noch über die Lösung von Handhabungs- und Sicherheitsproblemen, die Endverwertung und Rückführung in den Kreislauf der Natur hinaus. Diese Verantwortung sieht in der Natur weder ein Ausbeutungsobjekt noch einen Menschenfeind in ihren Gewalten, die oft erst in Reaktionen auf Menschenverhalten zur Wirkung kommen. Dafür muss man nicht Zeit und Ewigkeit bemühen, sondern im Jetzt und Hier einen sinnvollen Gebrauch seiner Gaben und Möglichkeiten finden. Gutes Design hilft, Selbstbewusstsein zu stärken, sich seiner selbst besser bewusst zu werden (ohne dass es nur für eine Elite ausgegrenzt wird). Um zu überzeugen, muss der Designer von guter Gestaltung auch genügend haben. Für kleine Hilfen, für intelligente Lösungen, für Aufforderungen zur Nachdenklichkeit im rasenden Zeitverbrauch aller Endlichkeiten.

As a producer, his studio does not stop at the product but creates the packaging, the sales tools, the advertising materials. As an agency man, he has marketing experience. As a freelancer, head designer and co-founder and board member of the small but excellent German Designer Club (DDC, 1989), he has solid, all-round professional practice. This is further bound with his activities not only as a product designer but also as a graphic, packaging, corporate, exhibition and architectural designer. The horizon of these professional spheres is therefore covered in full. His time at sea has given him panoramic vision. The design profession has freed itself from the narrowness of comfortable, prettifying and sales stimulating design, and has taken on a dimension of social responsibility. This even extends beyond the resolution of handling and safety issues, end use and bringing things back to the cycle of nature. This responsibility sees in nature neither an object for exploitation nor an enemy in her powers, which often only come into effect in response to human behaviour. For this, one must not worry about time and eternity, but find a use for one's talents and possibilities in the here and now. Good design helps strengthen self-confidence, self-awareness (without it becoming ringed off for an elite). In order to convince, the designer also needs to have enough good creations. For the little helpful things, intelligent solutions, demands for reflection in the rapid passage of time of all finiteness.

Design:
Design:
hinsehen und zuhören
look and listen
nachdenken und begreifen
consider and understand
ein Ziel setzen
set an objective
einsammeln und ausgrenzen
collect and limit
prüfen und vergleichen
check and compare
probieren, variieren,
try, vary,
selektieren, optimieren,
select, optimize,
reduzieren, realisieren
reduce, realize
erneut:
again:
hinsehen und zuhören
look and listen
nachdenken und begreifen
consider and understand
ein Ziel setzen.
set an objective.

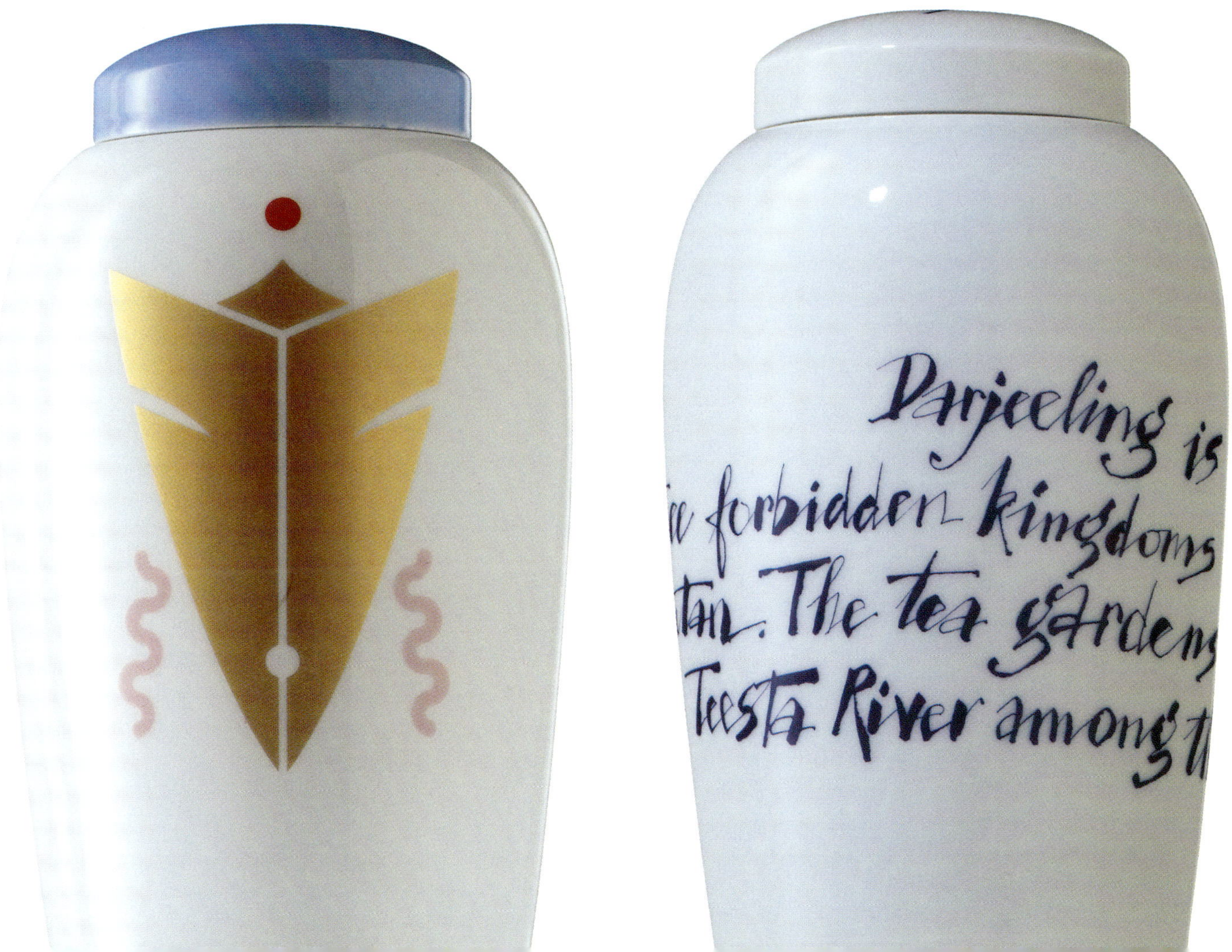

KOCHLABOR UND GAUMENSCHMAUS

500 g schwarze Spaghetti (die mit der Farbe des Tintenfisches gefärbt wurden)

1 Zwiebel

4 Knoblauchzehen

2 EL Olivenöl

20 g Butter

1 Döschen geriebener Safran

1 Döschen Faden-Safran

250 g Crème fraîche

Salz

Pfeffer

1 Glas trockener Weißwein (den gleichen, den man auch zum Gericht trinkt)

1 Würfel Fischfond

SEPIA-SPAGHETTI MIT SAFRAN-CREME

1. Spaghetti in reichlich Salzwasser nach Packungsanweisung al dente kochen, in ein Sieb abgießen und abtropfen lassen.

2. Für die Sauce Zwiebel und Knoblauch schälen, in kleine Würfel schneiden und zusammen mit dem Olivenöl und der Butter in eine Pfanne geben und glasig dünsten.

3. Den geriebenen und Faden-Safran in einer Schüssel mit der Crème fraîche gut verrühren und etwas Salz und Pfeffer zugeben sowie den Wein und den Fischfond. Die Masse in der Pfanne kurz umrühren und einmal aufkochen.

4. Die gekochten Spaghetti auf die Teller verteilen und mit der orange-gelben Sauce überziehen. Ein Vernaccia passt hervorragend dazu.

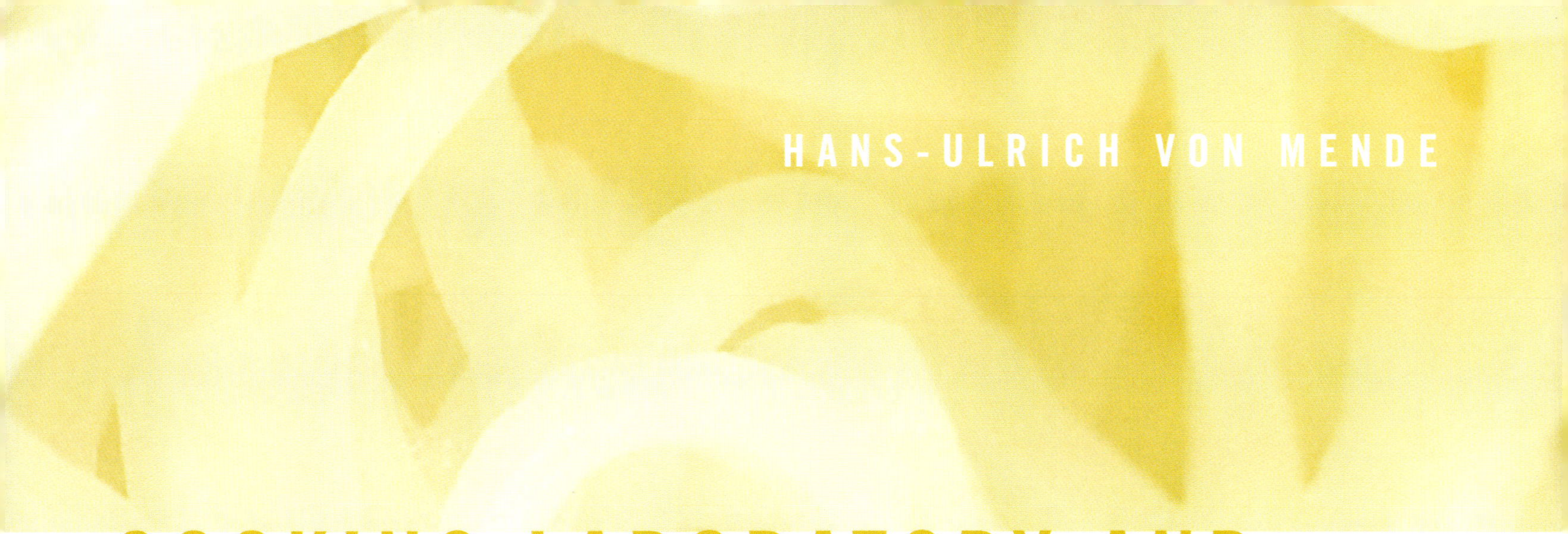

HANS-ULRICH VON MENDE

COOKING LABORATORY AND FEAST FOR THE TASTE BUDS

500 g black spaghetti
(dyed with colour from cuttlefish)
1 onion
4 cloves of garlic
2 tbsp olive oil
20 g butter
1 small tin ground saffron
1 small tin saffron strands
250 g crème fraîche
Salt
Pepper
1 glass dry white wine
(the same that will be drunk with the meal)
1 cube fish stock

SEPIA-SPAGHETTI WITH SAFFRON CREAM

1 *Boil the spaghetti in plenty of salted water according to the instructions on the packet until al dente, pour into a sieve and allow to drain.*

2 *Peel and dice the onion and garlic for the sauce, put into a pan with the olive oil and butter and fry until transparent.*

3 *Mix the ground saffron and saffron strands well with the crème fraîche and add some salt and pepper as well as the wine and the fish stock. Give the mixture a quick stir in the pan and bring to the boil once.*

4 *Divide the cooked spaghetti onto plates and cover with the orange-yellow sauce. A Vernaccia goes excellently with this dish.*

Weiße Haare machen alt und Designer tragen schwarz, so die gängige Meinung. Bei Tassilo von Grolman stimmt beides nicht. Seine weißen Haare wirken, als seien sie blond, was ihn jünger macht, als er ist. Und das mit dem schwarz, nun, Tassilo von Grolman, kurz TvG genannt, ist für mich einer der wenigen Männer, die farbiges Outfit lieben. Er beherrscht die Kombination von Farben wie kaum ein anderer und hätte auch der Boss von „Boss" werden können, wenn „Boss" mehr Mut zur Farbe hätte. Richtig bunt aber wird es erst, wenn TvG seinem Hobby frönt. Wer ihm beim Kochen zusieht, und das geht prächtig in seiner von ihm selbst eingerichteten Küche, sieht an Zutaten alle Farben dieser Welt. Tassilo von Grolmans Küchendesign ist dabei ebenso Werkstatt wie Designobjekt. Eine Werkstatt muss es sein, weil er perfekt kochen kann. Und echte Handwerker, wie gute Köche es sind, legen Wert auf Funktionalität. Die Form wird dann schon eine gute sein. Designambiente ist seine Küche natürlich auch, doch nur so weit, wie er sein Interesse an Design beweisen will. Wenn also TvG zur Arbeit, nein, zur Tat schreitet und das Ziel das Kochen ist, dann bleibt er ganz er selbst, er verkleidet sich nicht. Weiße Schürze und aufgekrempelte Ärmel genügen. Er gibt das typische Bild eines Menschen ab, der gerne anpackt, übrigens einer der charakteristischen Wesenszüge von Grolmans. Schon auf seinem Weg zum Design war Anpacken, also Praxis üben, sein Leben. Engländer nennen so etwas „learning by doing". Auch wenn es mit Umwegen zum Design ging, Tassilo von Grolman hat so viel im und vom Leben gelernt, dass er eigentlich herzlich wenig Theorie braucht, um seine Arbeit zu tun: Design machen oder eben kochen.

WIE GENIESSEN DESIGN BEEINFLUSST

Betrachten wir seine Entwürfe, haben alle etwas mit Genuss und Genießen zu tun. Auf den ersten Blick sieht man vielleicht keine Generallinie, kein Stilcredo, unter das alles unterzuordnen ist. Dennoch: Tassilo von Grolman hat eine Handschrift, die vom Leben geprägt ist, und auch nicht wenig vom Leben am Herd. Da zeigen sich seine Objekte voller Sinn. Wer kommt schon wie er auf die Idee eines kannengroßen Teesiebes. Er weiß eben aus sinnenfroher Erfahrung, wie Tee schmeckt, wenn er im Wasser schwimmen darf. Typisch für TvG auch sein Brotkasten-Design: Kein Metall, sondern Holz, die Krümel unter einen Rost verbannt. Und das Brotmesser sucht man sowieso immer in der falschen Schublade. Darum einfach eine Ablagemulde mit definiertem Messerumriss. Um zu verhindern, dass sich beim Öffnen die Finger an den Bügelgriffen klemmen, kam von Grolman auf einen einfachen Trick: Hauben- und Brettbügel liegen nicht designerschick parallel übereinander, sondern sind einmal nach innen und einmal nach außen geschweift. Das ist wahres TvG-Design. Wenn dieses wunderbare Stück Holz dann noch „Krümelmonster" heißt, dann hat sich der Designer mit Humor – oder war es der Schalk? – in diesem Objekt verewigt.

Linke Seite
Auf gute Zutaten kommt es an. Tassilo von Grolman, 1986.

Rechte Seite
Tassilo von Grolman in seiner Küche, 2000.

Left-hand side
So much depends on good ingredients. Tassilo von Grolman, 1986.

Right-hand side
Tassilo von Grolman in his kitchen, 2000.

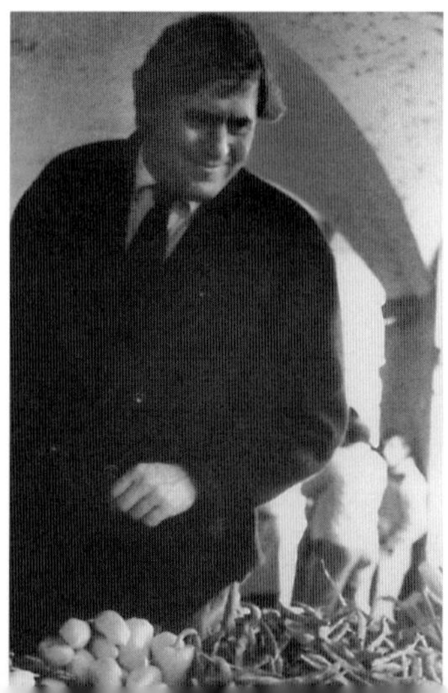

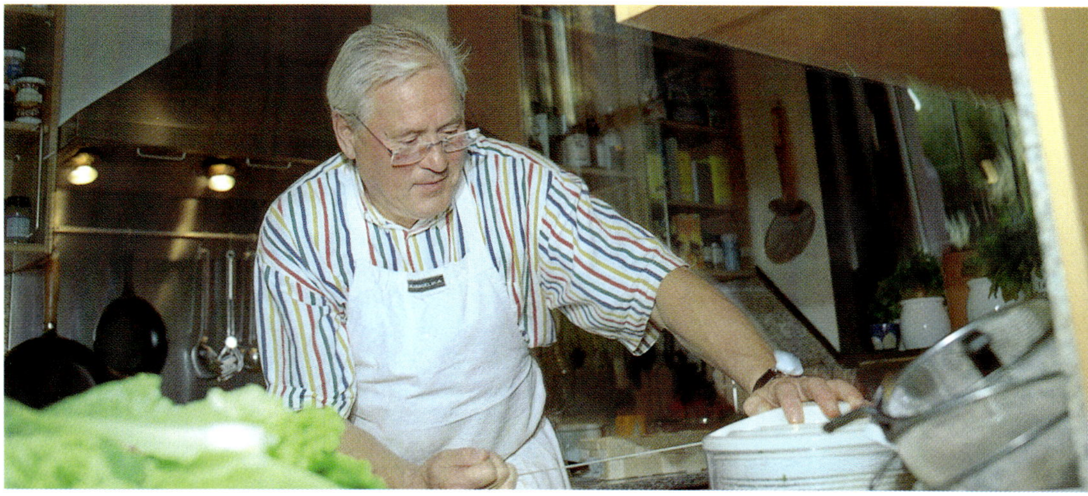

White hair makes you look old and designers wear black, according to general opinion. Neither is true of Tassilo von Grolman. His white hair appears to be blond, making him look younger than he is. And the point about wearing black, well, Tassilo von Grolman – known as TvG for short – is, to me, one of the few men who loves colourful clothes. He masters the combination of colours as almost no other, and he could have been the boss of "Boss", had "Boss" used colour more courageously. However things only get really colourful when TvG is indulging in his hobby. Whoever watches him cook – which is splendidly easy in the kitchen that he fitted out himself – sees all the colours of the world in the ingredients. The design of Tassilo von Grolman's kitchen is as much workshop as it is design object. It has to be a workshop, because he can cook perfectly. And real craftsmen, as good cooks are, value functionality. The design will then certainly be a good one. His kitchen is also, of course, a design environment but just enough to prove his interest in design. So when TvG gets down to work – no, to the action – and the objective is cooking, he is just himself, he does not disguise himself. A white apron and rolled-up sleeves suffice. He conveys the typical image of a man who likes to get stuck in, which, by the way, is one of von Grolman's defining traits. Even on his route to design, getting stuck in, or exercising practical application, was his life. The English call this "learning by doing." Even if he got to design by a detour, Tassilo von Grolman has learnt so much in and from life that he actually needs precious little theory to do his work: whether designing or cooking.

HOW ENJOYMENT INFLUENCES DESIGN

If we look at his designs, they all have something to do with pleasure and enjoyment. At first sight there is perhaps no general tendency or style credo, which subordinates everything. However: Tassilo von Grolman's handwriting is shaped by life and, not least, by life in the kitchen. His objects appear to be full of sense. Who else but he would come up with the idea of a tea strainer that is as big as the teapot. He just knows from pleasurable experience how tea tastes if it is allowed to swim in the water. His bread bin design is also typical of TvG: not metal but wood banishes the crumbs under a grille. And the bread knife can never be found anyway because it is in the wrong drawer. So there is simply a storage trough with a defined knife shape. To avoid the problem of trapping fingers in the handles on opening, von Grolman discovered a simple trick: the lid and board handles are not parallel to each other in designer style but are curved once to the inside and once to the outside. That is true TvG design. When this wonderful piece of wood is then called "Krümelmonster" (crumb monster), then the designer has immortalised himself with humour – or was it mischief? – in this object.

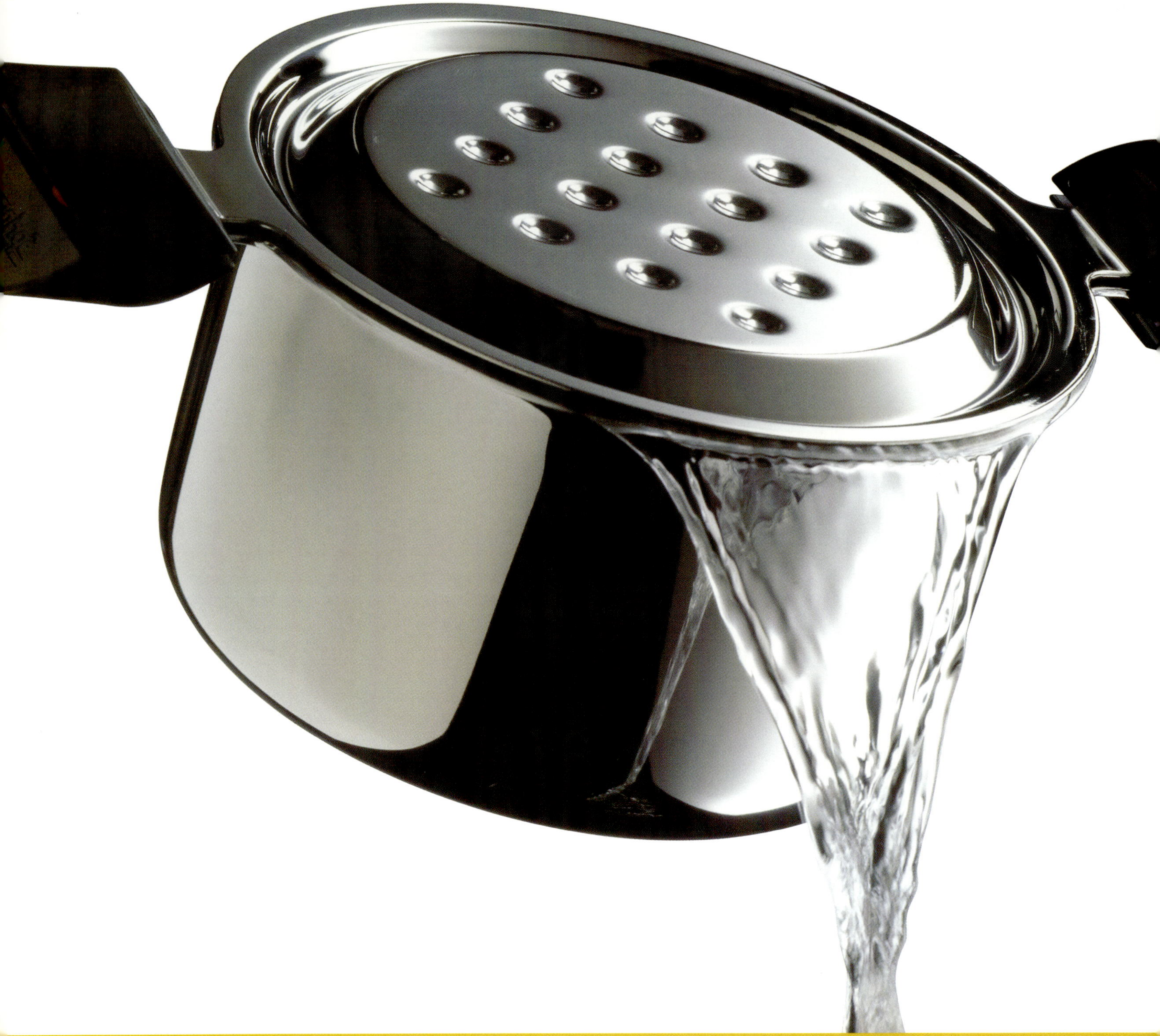

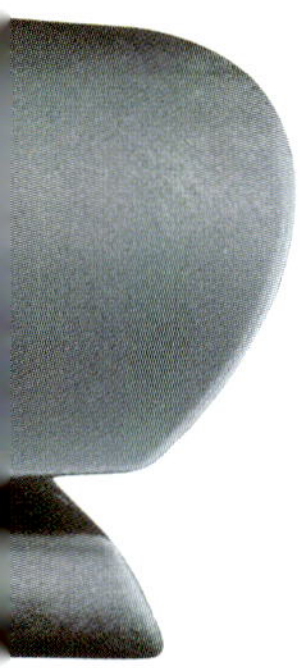

Diese und folgende Seite
Kochtopfserie „Compact GLI“ für
die Schweizer Spring AG, 1992.

Above and overleaf
Cooking pan series “Compact GLI”
for the Swiss Spring AG, 1993.

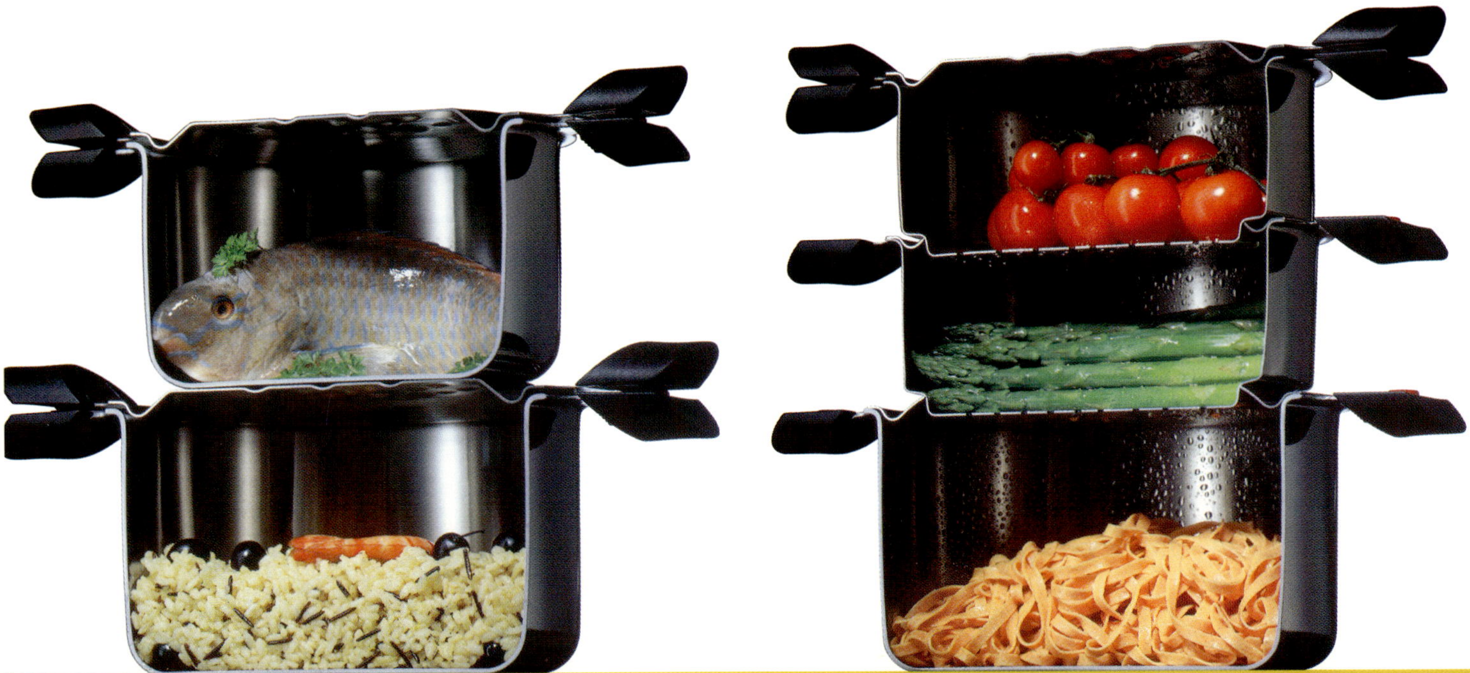

ENTWURFSSCHMIEDE KÜCHE

Mehr noch als für Teetrinken und Brotschneiden steht Tassilo von Grolman für absoluten Kochgenuss. Da ist es kein Wunder, dass der Koch von Grolman, bei dem die Essen immer mehrgängig sind, dem Designer von Grolman gleich eine ganze Topfserie aufs Papier diktierte. Weil ein Koch in voller Arbeit immer viel zu köcheln hat, muss er höllisch aufpassen, dass ihm nichts anbrennt, verkocht oder unter der Hand zerfällt. Tempo ist gefragt und Tempo verlangt kontrolliertes Handeln. Wer in der Hektik danebengreift und sich verbrannte Finger holt, hat womöglich falsche Töpfe mit falschen Griffen. Wenn dazu noch möglichst viel gleichzeitig zu garen ist und der Herd schon zugestellt ist mit Pfannen und Töpfen, dann ist die Idee nicht weit, Töpfe mit flachen Deckeln und seitlichen Griffen zu kreieren, so dass man sie während des Kochvorgangs stapeln kann. Dieses Kochen in Etagen ermöglicht, unten das Essen mit größtem Hitzebedarf, oben das schonende, temperierte zu garen. Wo andere Restwärmenutzung als das ökologisch korrekte Nonplusultra verkaufen, macht's von Grolman ohne viel Tamtam, einfach so, wie es einst Hausmütter machten, als die Herde winzig und die Kohlen knapp waren.

KITCHEN AS A DESIGN FORGE

Tassilo von Grolman stands for much more than drinking tea and cutting bread – he stands for absolute pleasure in cooking. So it is no wonder that the cook von Grolman – whose meals always feature several courses – dictated a whole set of saucepans at once onto paper for the designer von Grolman. Because a cook at work always has many things on the go, he has to be desperately careful that nothing burns, boils away or falls apart while in his care. Speed is of the essence and speed demands controlled actions. Whoever reaches wrongly for something in the hectic rush and burns his fingers possibly has the wrong pans with the wrong handles. When, on top of that, as many things as possible have to be cooked at the same time and the hob is already full of pots and pans, it is a short step to the idea of creating pans with flat lids and handles on the side so that they can be stacked during cooking. This cooking on different levels makes it possible to cook the food that needs the most heat underneath and that which needs to simmer gently at the right temperature above. Where others sell the use of residual heat as the ecologically correct ultimate, von Grolman does it without much fuss – simply in the way that housewives used to do it, when the ovens were tiny and coal was scarce.

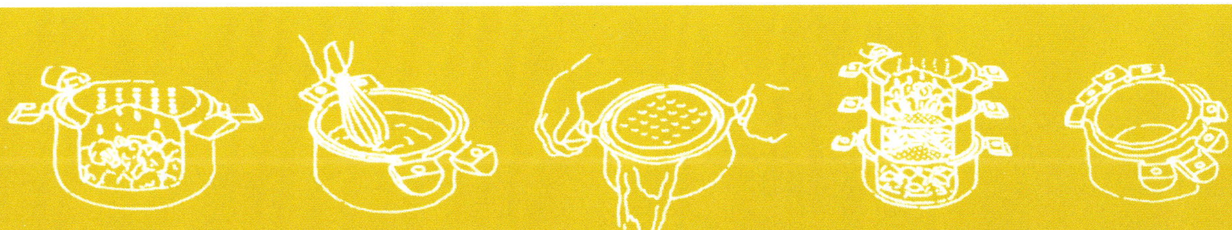

CHANGE

Design Tassilo von Grolman

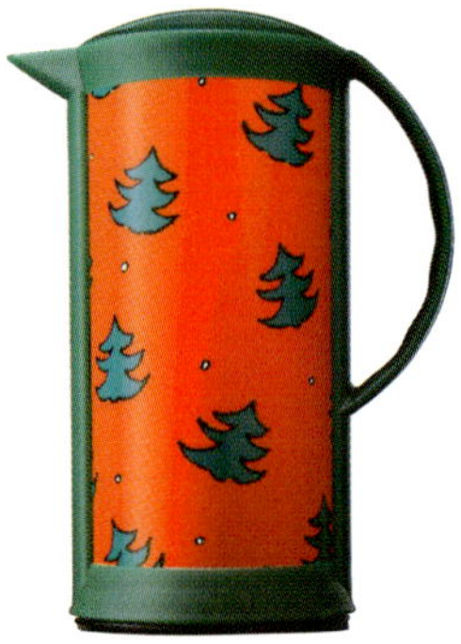

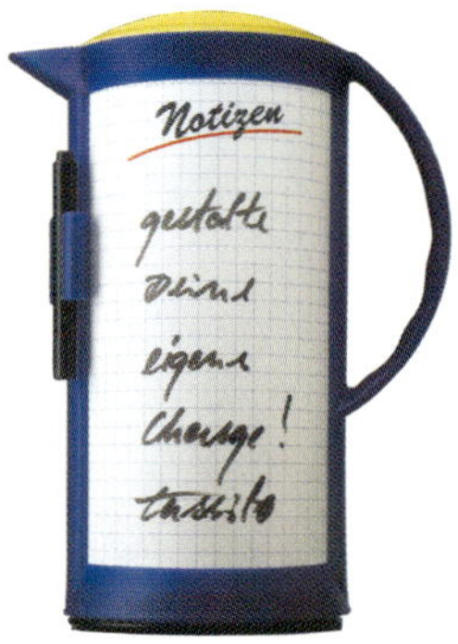

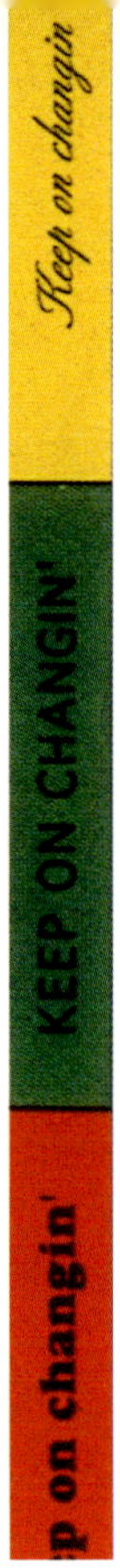

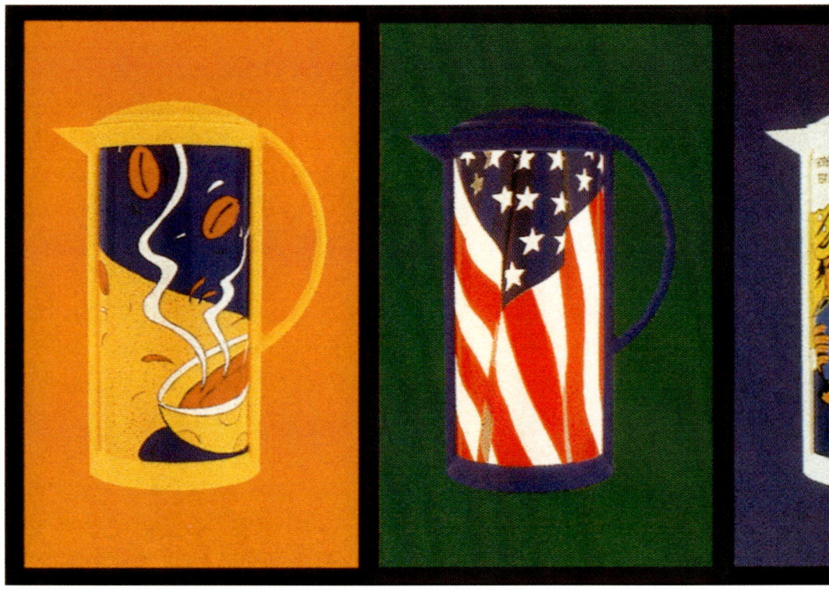

Wenn er dann zu Tisch bittet, hat Tassilo von Grolman nicht vergessen, dass seine Gäste auch mit dem Auge essen. Seine Tische sind stilvoll gedeckt, kein Löffel, der einem zugunsten der Optik die Suppe an den Mundwinkeln runterrinnen lässt, keine Gabel, die eher einer Form als einer Funktion folgt, kein Messer, das den druckausübenden Finger abrutschen lässt, nur weil es zum Komplett-Besteckdesign passen muss. Kurz gesagt, es ist eine Freude, bei ihm zu essen. Das Bier in einem von ihm gestalteten Glas perlt fröhlicher, und wenn er will, bringt er zum Abschluss den Kaffee in seiner Thermoskanne „Change", die ihr Kleid täglich wechseln kann oder das Wachhalte-Getränk mit witzigen Sprüchen umspannt, solange es noch in der Kanne ist. Darf ein Designer so eine Spaßkanne eigentlich machen? Aber natürlich, eine Thermoskanne ist doch nicht der Gipfel der Designerwelt! Da darf schon mal ein bedruckter Kunststoffzylinder eingeschoben werden, um den 365 Tagen des Jahres 365 Thermoskannen zu gönnen.

When inviting them to the table, von Grolman has not forgotten that his guests also eat with their eyes. His tables are laid with great style; no spoon, which look great but make the soup run down from the corner of the mouth; no fork which has more design than function; no knife which lets the finger slip down when pressure is applied just because it has to fit the shape of the whole cutlery set. In short, it is a joy to eat with him. The beer, in a glass designed by him, effervesces more cheerfully and when he chooses to he brings in the coffee in his thermos flask "Change", which can change its dress daily or can transform the keep-awake drink with little witty sayings for as long as it is still in the pot. Should a designer really make such a fun coffee pot? Of course he should – a thermos flask is not the pinnacle of the world of design! A printed plastic cylinder can certainly be squeezed in to allow 365 thermos flasks for the 365 days of the year.

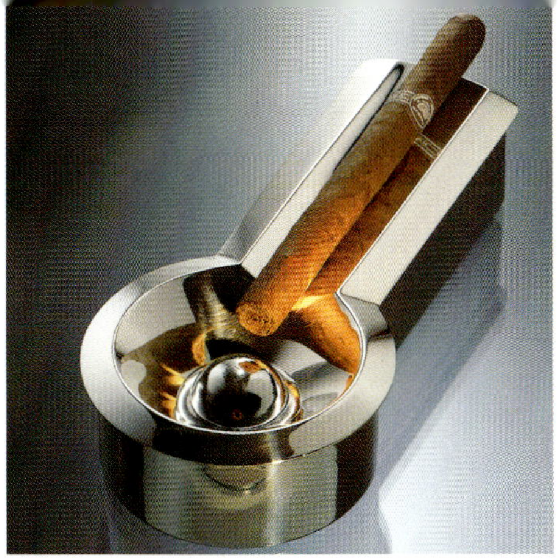

Linke Seite
„Change" Isolierkannen für Alfi, 1994.
Die Gestaltungsmöglichkeiten der Isolierkanne sind nahezu grenzenlos. Eine Variante der „Change" kann als Untergrund für Notizen verwendet werden, die Weihnachtskanne überrascht beim Eingießen mit einem Lied.

Rechte Seite
Zigarrenaschenbecher „Hommage à Marianne Brandt". Studie in Silber, 1990.

Left-hand side
"Change" thermal pots for Alfi, 1994. The design possibilities of the thermal pots are nearly endless. One variant of the "Change" can be used as a surface for notes, the Christmas pot surprises us with a song when pouring.

Right-hand side
Cigar ashtray "Hommage à Marianne Brandt". Study in Silver, 1990.

ZIGARREN UND KOCHBÜCHER FÜRS LEBEN

Tassilo von Grolman wäre kein Genießer, wenn zum Essensabschluss nicht auch geraucht würde. Wer Zigaretten bevorzugt, darf seine Asche in die Kreismulde inmitten eines weißen Quadrats versenken. Die Ablagefläche für den Glimmstengel ist das negative Abbild seiner selbst und findet seine positive Entsprechung am Aschenbecherboden als halbrunden Steg. So gestaltet bildet die Fläche Füße, die weder Kratzspuren auf edlem Tischholz hinterlassen, noch den Becher kippeln lassen. Tassilo von Grolman wird nach dem Essen jedoch zur Zigarre greifen, was eigentlich unvorstellbar bei diesem rastlosen Mann ist. Und doch passt die Zigarre perfekt zu ihm, der vor allem Genießer ist. Perfekt, meint er, sollte auch der Zigarrenascher sein. Weil ihm die Pause beim Rauchen wichtig ist, vermisste er immer eine sinnvolle Ablage für das teure Kraut, eine, die die Asche gezielt auffangen kann. Da er keinen geeigneten Zigarrenascher fand, hat er ihn kurzerhand selbst geschaffen. Nun hat er beides: die Pause beim Rauchen und den passenden Ascher dazu. Aber hätten wir das erzählen dürfen, wo er als perfekter Koch das Rauchen doch ablehnen müsste, weil es die Geschmacksnerven malträtiert? Aber der Koch wie auch der Designer von Grolman ist nicht dogmatisch. Da zählt Genuss, nicht Credo. Für ihn sind auch Bauhaus und Ulmer Schule mitnichten die einzig wahren Philosophien, wie man gestalten soll. Von Grolman liest in diesen Lehren eher wie im Kochbuch: Man nehme! (damit Menge und Garzeit stimmen). Wie es schmecken wird, kommt letztlich immer noch auf den Koch an!

CIGARS AND COOKERY BOOKS FOR LIFE

Tassilo von Grolman would not be a pleasure-lover, if the meal did not also conclude with a smoke. Those who prefer cigarettes can push their ash into the circular hollow in the centre of a white square. The place for leaving the cigarettes is the negative image of itself and finds its positive equivalent at the bottom of the ashtray as a semicircular ridge. Designed in this way, the piece forms feet that will neither leave scratches on precious table wood nor allow the ashtray to tip. Tassilo von Grolman, however, will reach for a cigar after the meal, something that is really unimaginable for this restless man. And yet the cigar suits him perfectly as, above all, he is a connoisseur. The cigar ashtray, he says, should also be perfect. As the pauses are important to him when smoking, he had always missed a sensible place to put the expensive weed – one that can specifically catch the ash. As he found no suitable cigar ashtrays, he made one himself without further ado. Now he had both: the pauses while smoking and the ashtray to match. But should we have talked about this, when as a perfect cook he should decline smoking because it maltreats the taste buds? But the cook, just like the designer von Grolman, is not dogmatic. Enjoyment is what counts, not credo. For him even Bauhaus and the Ulm school are by no means the only philosophies that dictate design. Von Grolman reads these teachings just as he would a cookery book: Take! (so that quantities and cooking times are correct). In the end, the taste still depends on the cook!

Produkte haben für Tassilo von Grolman auch eine psychologische Dimension. Um auf den emotionalen Aspekt von Design hinzuweisen, beauftragte er die New Yorker Fotokünstlerin Anja Conrad, Menschen zu portraitieren, die ihm nahe stehen und seine Produkte benutzen. Anja Conrad studierte in Chicago und New York. Als freiberufliche Fotografin arbeitet sie heute sowohl in den USA als auch in Europa.

For Tassilo von Grolman, products also have a psychological dimension. In order to show the emotional aspect of design, he commissioned the New York photo-artist Anja Conrad to create portraits of people who are close to him and who use his products. Anja Conrad studied in Chicago and New York. Today she works as a freelance photographer in the USA as well as in Europe.

Das Interview führte Katja Kretzschmar im September 2000./*Katja Kretzschmar conducted this interview in September 2000.*

ZWISCHEN EMOTION UND RATIO
BETWEEN EMOTION AND REASON

Interview

Herr von Grolman, Ihre Lehre absolvierten Sie bei der AEG, danach studierten Sie Maschinenbau und schließlich Industrial Design in Kassel. Doch Ihr Traum war es, in Ulm an der Hochschule für Gestaltung zu studieren.
Bis dahin war es ein weiter Weg. Ich war ein schlechter Schüler. Mein Onkel vermittelte mir eine Lehrstelle zum Maschinenschlosser bei der AEG. Elektronik interessierte mich schon immer und ich bastelte gerne. Während meiner Lehre wurde mir klar, dass ich Designer werden wollte. Also habe ich der AEG viel zu verdanken. Die Zeit dort hat mir Glück gebracht. Danach begann ich, Maschinenbau zu studieren – auf Wunsch meines Vaters. Ich hatte keine Lust, Ingenieur zu werden und feierte lieber bis morgens Drei. Das Studium kam zu kurz und prompt schaffte ich das Examen nicht.

Und dann fuhren Sie auf der Santa Rosa zur See. Wie kam es dazu?
Ich sollte zur Bundeswehr, auf die ich erst recht keine Lust hatte. Ich fand heraus, dass nach § 5 des Wehrpassgesetzes diejenigen von der Bundeswehr befreit sind, die zur christlichen Seefahrt gehen. Ich bin dann nach Hamburg gefahren. Ich wollte nach Amerika und habe auf der Santa Rosa angeheuert. Auf dieser Seefahrt hatte ich dann sehr viel Zeit zum Nachdenken. Und einen Satz gelernt: geht nicht, gibt's nicht ...

Doch Sie schlossen trotzdem Ihr Maschinenbaustudium ab. Obwohl Sie nicht als Ingenieur arbeiten wollten. Woher nahmen Sie die Ausdauer?
Vielleicht die Gene meiner preussischen Vorfahren? Nein, ich wollte eine klare Vita haben. So ging ich nach Berlin und beendete mein Studium, um dann endlich meinen Traum wahr zu machen und in Ulm zu studieren. Das war Anfang 1967. Doch die Hochschule für Gestaltung in Ulm sollte gerade geschlossen werden.

Herr von Grolman, you completed your apprenticeship with AEG, then you studied engineering and eventually industrial design in Kassel. But it was your dream to study at the Hochschule für Gestaltung (College of Design) in Ulm.
It was a long way there. I was a bad student. My uncle arranged an apprenticeship for me as an engine fitter with AEG. I still find electronics interesting and I liked doing things with my hands. During my apprenticeship I realised that I wanted to become a designer. So I have much to thank AEG for. The time there has brought me good fortune. After that I began to study engineering – following the wishes of my father. I did not want to become an engineer and preferred to party until three in the morning. The studying suffered and I promptly failed the exam.

And then you went to sea on the Santa Rosa. How did that come about?
I was supposed to do my national service in the army, which I really did not want to do. I found out that, according to section 5 of the German service record law, those who go on a Christian sea journey are absolved from national service with the army. I then drove to Hamburg. I wanted to go to America and signed up with the Santa Rosa. I then had plenty of time to reflect on the sea journey. And learnt one thing: if you don't try, you don't get ...

But you still completed your engineering studies. Although you did not want to work as an engineer. Where did you get the stamina from?
Maybe from the genes of my Prussian ancestors? No, I wanted to have a clear CV. So I went to Berlin and finished my studies so that I could then finally make my dream come true and study in Ulm. That was the beginning of 1967. But the Hochschule für Gestaltung (College for Design) in Ulm was just about to be closed.

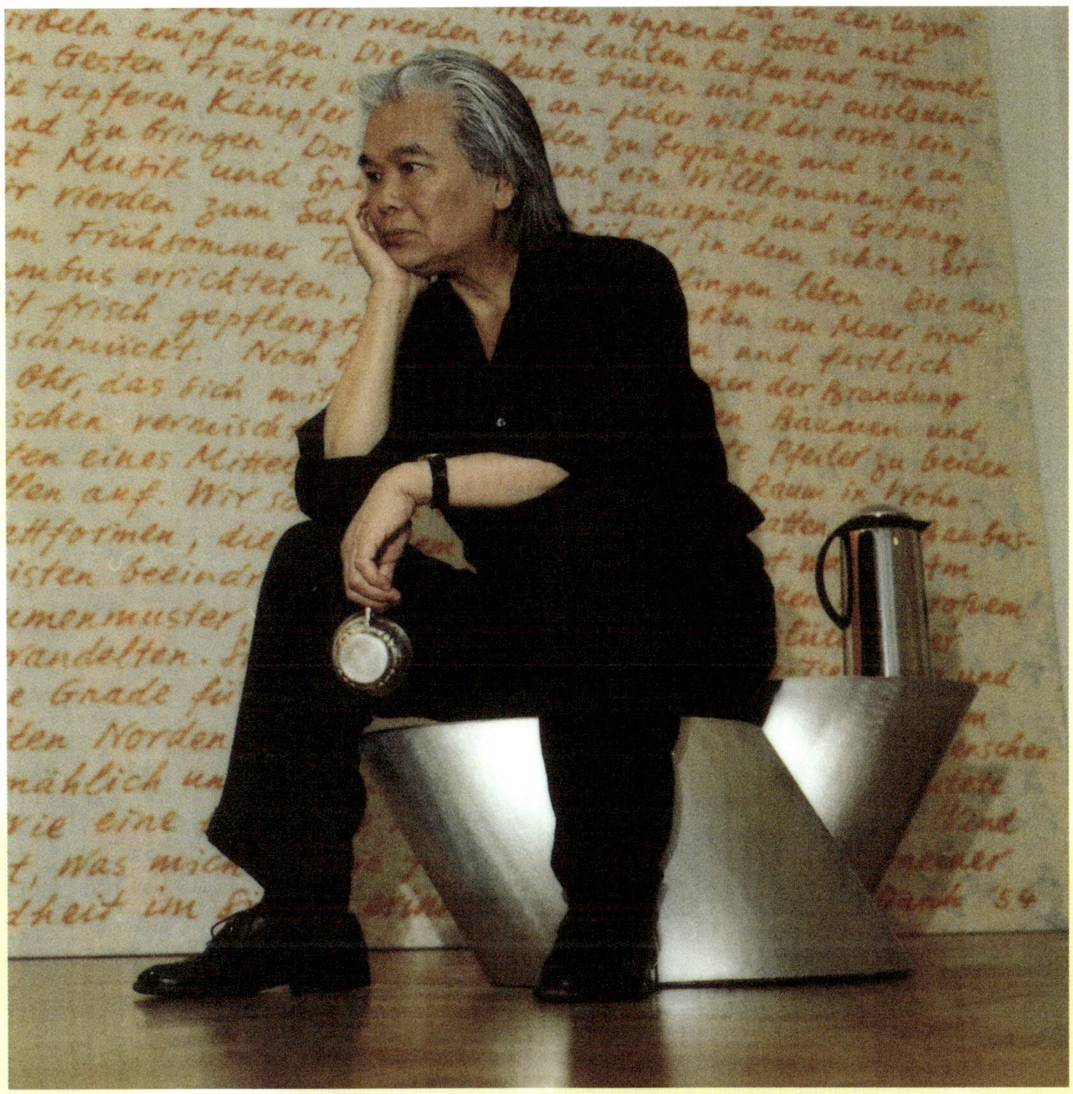

Oanh Pham-Phu
Bildhauer, Designer, Grafiker
Sculptor, Designer, Graphic Artist

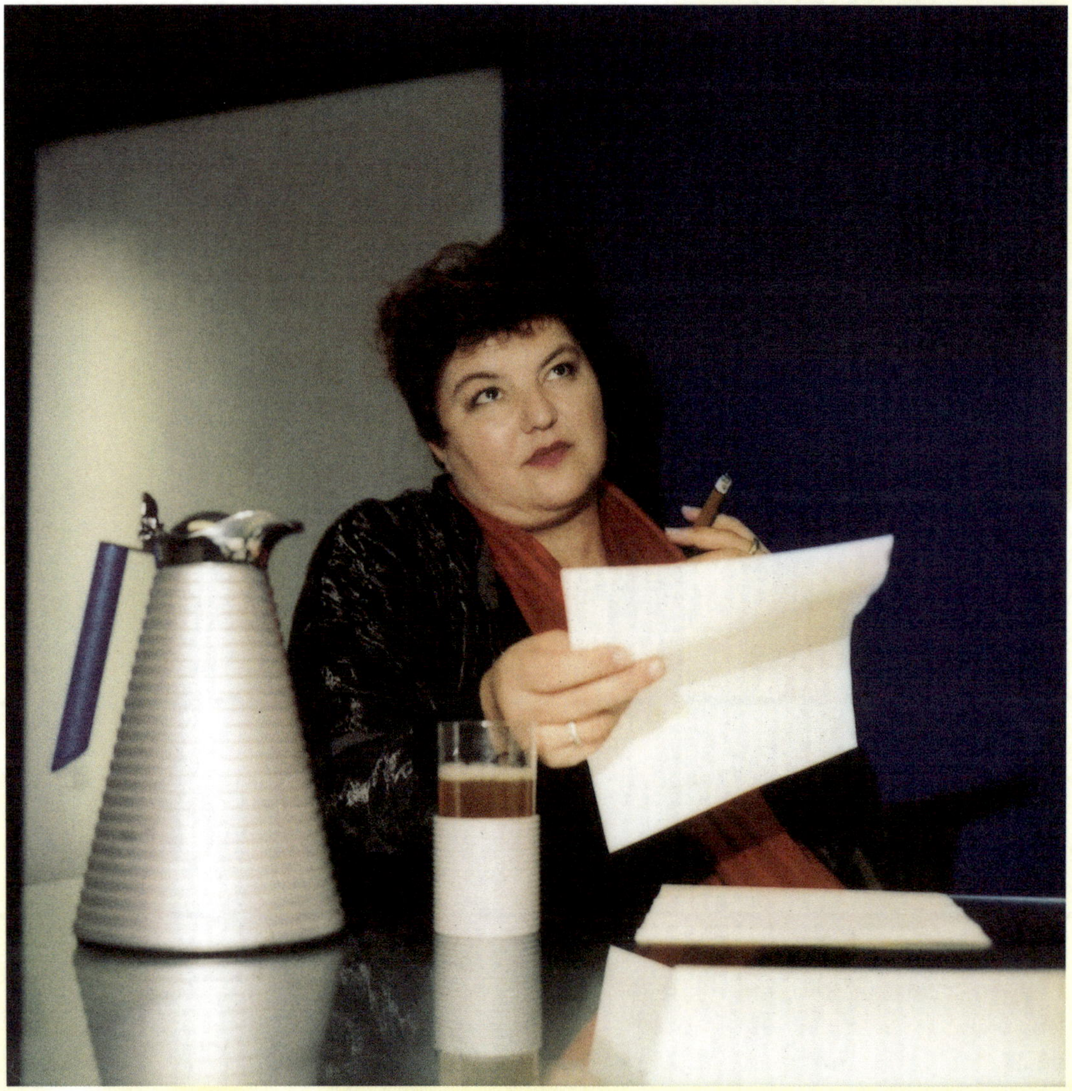

Bibs Hosak-Robb
Designerin
Designer

Was hat Sie an Ulm fasziniert?
Ulm war Geisteshaltung. Wollte man zu dieser Zeit Designer werden, ging man nach Ulm. Wir waren vom Existentialismus beleckt – Sartre und Camus – und liefen herum wie Beatniks mit Parka. Ulm war eine Beatnik-Schule für mich. Heute bin ich allerdings nicht mehr „Ulm-minded". Ich denke, dass dort der Versuch gemacht wurde, Emotionen zu verwissenschaftlichen. Emotionen sind aber nicht wissenschaftlich zu erklären. Emotionen sind Emotionen. Ich habe weiß Gott nichts gegen Otl Aicher, er hat großes geleistet, ist mir aber in vielem zu trocken. Dabei waren die „Ulmer" an sich lustvoll, hatten Sinn für Humor und für Sex. Aber diese Manie, alles reduzieren zu müssen! Minimalismus bis zum Erbrechen. Ganz anders das Bauhaus, das mich heute viel mehr fasziniert als Ulm.

Sie studierten kurze Zeit in Berlin an der Akademie für Werkkunst und Mode. Und dann ging es weiter an die Werkkunstschule Kassel, später Hochschule für Bildende Künste. Wie hat diese Zeit Sie geprägt?
In Berlin ging die Post ab. Die Studenten revoltierten, und das war für mich natürlich bombastisch. Die Akademie war mir allerdings für meine Ideen nicht aufgeschlossen genug. In Kassel studierte ich bei Prof. Herbert Oestreich und Prof. Günter Kupetz. Der damalige Direktor Jupp Ernst ermöglichte mir, zusätzlich sowohl Kurse in Architektur, Grafik und Foto zu belegen. Für ein Jahr war ich Assistent bei Professor Arnold Bode, dem Vater der documenta. Er war Vorbild, Lehrer und Genie zugleich.

Wer gehört noch zu Ihren Vorbildern?
Es gibt eine ganze Menge Vorbilder. Ich liebe zum Beispiel Anna Castelli-Ferrieri, die Grande Dame des Design, die mit 78 Jahren immer noch arbeitet. Ettore Sottsass, der große Philosoph, für mich sicherlich einer der größten Designer. Terence Conran, der lebensbejahende Gestalter. Ihm fühle ich

What fascinated you about Ulm?
Ulm was a state of mind. If you wanted to be a designer at that time, you went to Ulm. We were touched by existentialism – Sartre and Camus – and ran around like beatniks in parkas. Ulm was a beatnik school for me. Mind you, I am not so "Ulm-minded" today. I think that an attempt was made there to make a science out of emotions. Burt emotions can not be explained scientifically. Emotions are emotions. God knows, I have nothing against Otl Aicher – he has done some great things – but is too dry for me in many areas. Yet the people of Ulm usually were full of relish, had a sense of humour and a sense of sex. But this mania for reducing everything! Minimalism until it makes you sick. Bauhaus, which fascinates me today much more than Ulm, is utterly different.

You studied for a short time in Berlin at the Akademie für Werkkunst und Mode (Academy of arts, crafts and fashion). And the next stage was at the Werkkunstschule (arts and crafts college) in Kassel, which later became the Hochschule für Bildende Künste (College of Fine Arts). How did this time affect you?
Berlin hit eventful times. The students revolted and of course that was bombastic for me. Mind you, the academy was not open-minded enough for my ideas. I studied with Prof. Herbert Oestreich and Prof. Günter Kupetz in Kassel. The then director, Jupp Ernst, made it possible for me to take additional courses in architecture, graphic design and photography. I was an assistant to Professor Arnold Bode, the father of "documenta", for one year. He was role model, teacher and genius combined.

Who else do you consider to be your role models?
There are quite a lot of role models. For example, I love Anna Castelli-Ferrieri, the Grande Dame of design, who is still working at the age of 78. Ettore Sottsass, the great philosopher, for me certainly one of the greatest designers. Terence Conran,

mich in meiner Einstellung sehr nahe. Für mich ist Conran derjenige, der Design vorlebt. Auch Marianne Brandt vom Bauhaus ist ein Vorbild für mich. Ihr widmete ich den Aschenbecher „Hommage à Marianne Brandt".

Was bedeutet für Sie guter Geschmack und was hat Ihren Geschmack beeinflusst?

Geschmack kann man erben. Ich habe ihn von meiner Mutter geerbt, die sehr kreativ war. Geschmack lernen kann man vielleicht. Aber um stilsicher zu sein, muss man sehr viel Intuition haben und aus sich herausgehen können. Mit 30 sollte man ziemlich genau wissen, was gut ist und was schlecht oder mittelmäßig. Doch Geschmack bedeutet ja nicht nur, wie man sich kleidet oder wie man wohnt. Geschmack ist ja auch, wie der Tee, der Wein, das Essen schmeckt.

Sie sollen ein begnadeter Koch sein. Und viele Ihrer Produkte stammen aus dem Umfeld Küche und Tischkultur. Kommen Ihnen neue Ideen beim Kochen?

Ja. Zum Beispiel ist das Messer eine entscheidende Sache. Warum schneiden 80 Prozent aller deutschen Hausfrauen mit einem kleinen Kneipchen? Weil sie Angst haben, sich zu verletzen. In Asien dagegen arbeiten die Frauen mit großen Messern. Der Unterschied: Diese Messer sind nicht spitz, sondern vorne abgehackt. Da kam mir die Idee, ein Messer zu gestalten, dessen Spitze stumpf ist, das aber dennoch scharf genug ist, um damit schneiden zu können. Herausgekommen sind Messer mit runder Spitze, die ernst genommen werden sollten.

Lebensfreude und Sinnlichkeit sagt man Italien und dem italienischen Design nach. Sehen Sie das auch so?

Schon die Römer waren unheimlich kreativ. Gut, sie haben von den Griechen geklaut, aber sie haben sie weiterentwickelt, und das zählt doch! Die Italiener werden tagtäglich mit alter,

the life-affirming designer. I feel very close to him in my outlook. For me, Conran is one of those who live design. Marianne Brandt of Bauhaus is also a role model for me. I dedicated the ashtray "Hommage à Marianne Brandt" to her.

What does good taste mean for you and what has influenced your taste?

You can inherit taste. I inherited it from my mother, who was very creative. You can possibly learn taste. But to be sure of your style you have to have plenty of intuition and be able to come out of your shell. By the age of 30, you should know pretty well what is good and what is poor or mediocre. But then taste does not just mean how you dress or how you live. Taste is also appreciation of tea, wine, food.

You are reputed to be a talented cook. And many of your products stem from the area of the kitchen and the culture of the table. Do you get new ideas when you are cooking?

Yes. For example, the knife is a crucial thing. Why do 80% of all German housewives cut with a little knife? Because they are worried about injuring themselves. In Asia, by contrast, the women work with big knives. The difference: these knives are not pointed, but squared off at the end. So I got the idea of designing a knife that has a blunt end but is still sharp enough to cut. From this came knives with a rounded end which should be taken seriously.

Italy and Italian design suggest joy of living and sensuality. Would you agree with that?

The Romans were certainly incredibly creative. OK, so they stole from the Greeks, but they developed things further and that counts! The Italians are confronted with old, good architecture day in, day out. Just think of Florence or Venice or Milan. Excellent taste – and so good style – could develop from this. If only we also still had intact architecture, perhaps

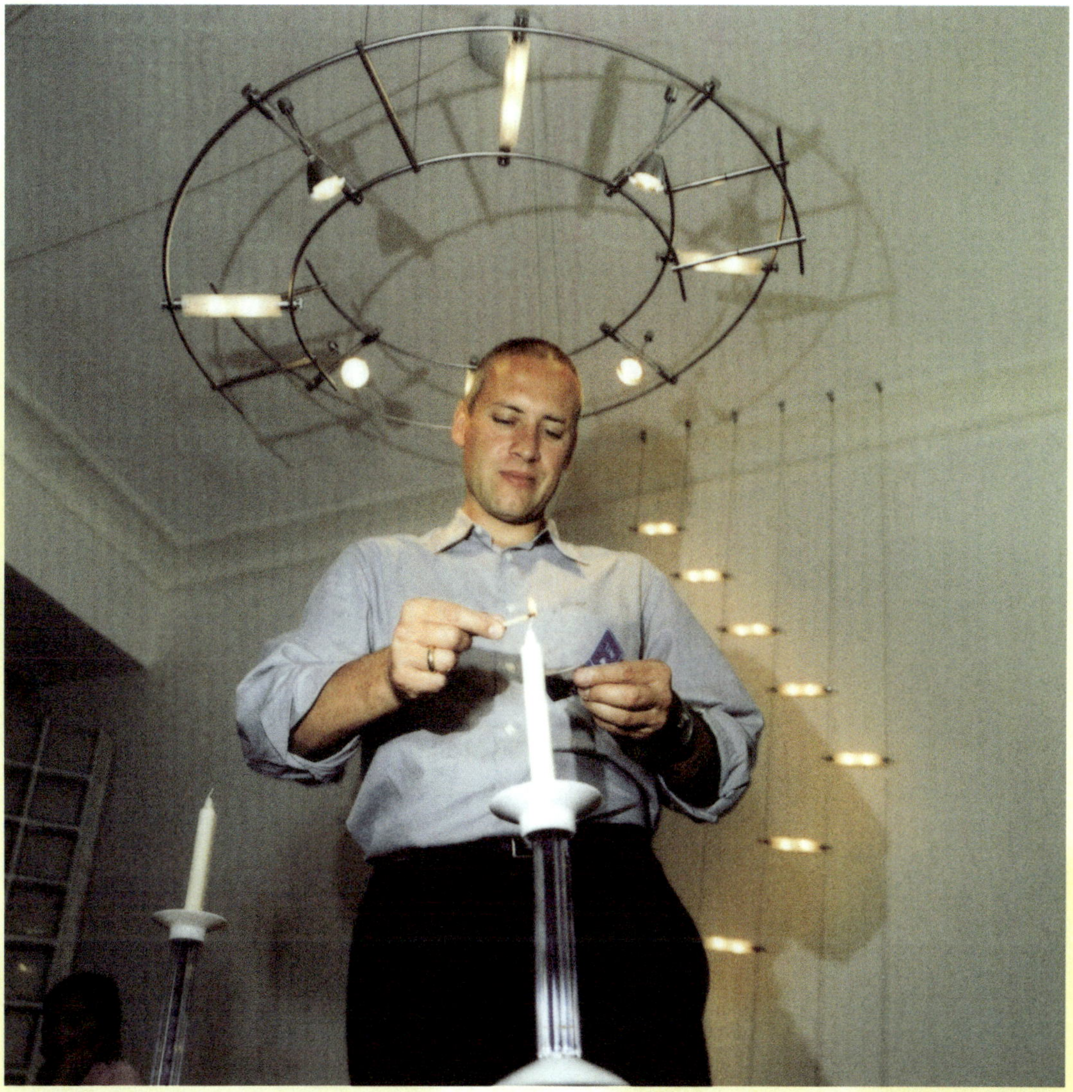

Axel Meise
Licht-Designer
Light Designer

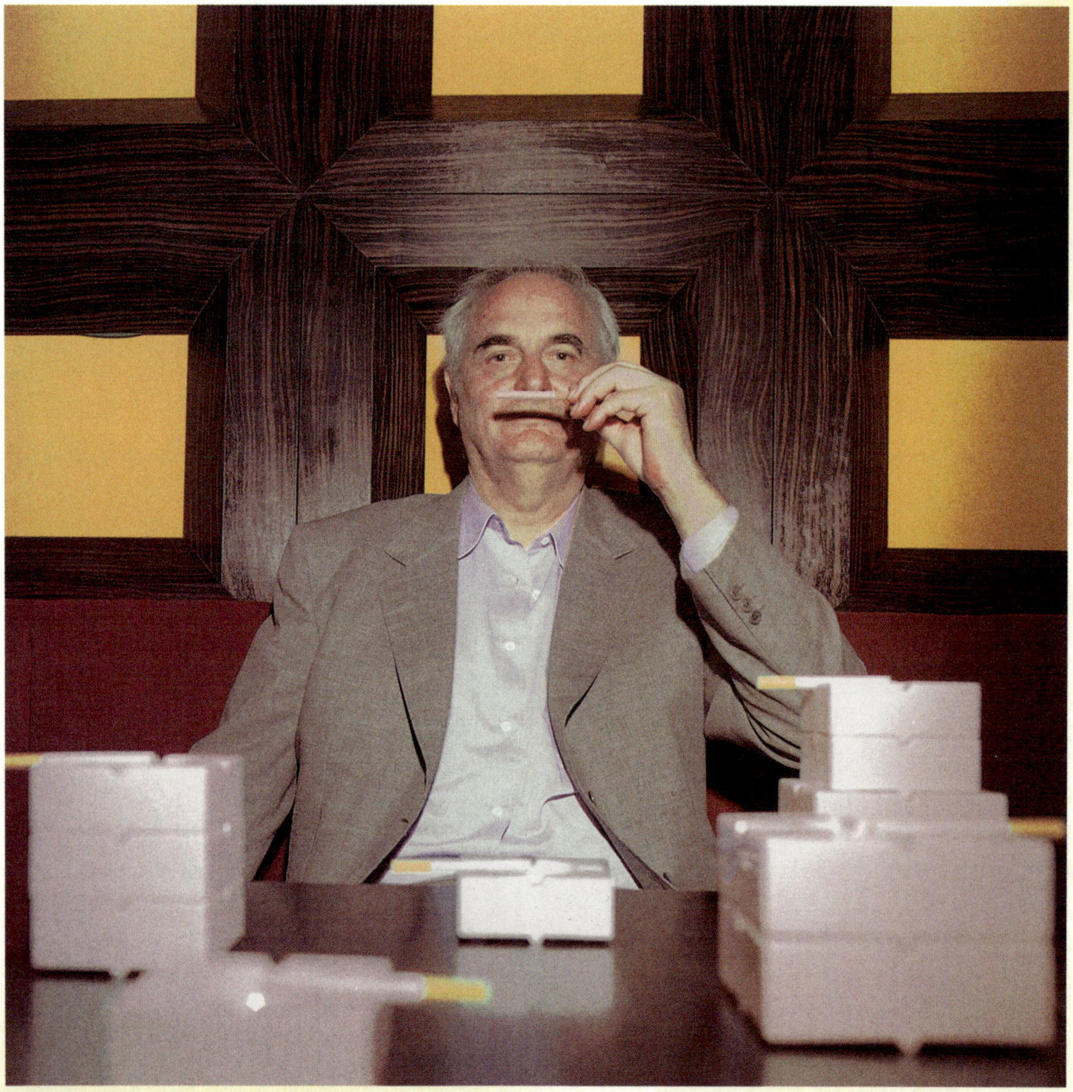

Michael Conrad
Creative Chief Officer Leo Burnett Company Inc.
Creative Chief Officer Leo Burnett Company Inc.

guter Architektur konfrontiert. Denken Sie nur an Florenz oder Venedig oder Mailand. Daraus konnte sich ausgezeichneter Geschmack, also ein guter Stil, entwickeln. Hätten wir ebenfalls noch intakte Architektur, dann wäre vielleicht auch in Deutschland die Lust am Design größer. Und dann essen die Italiener gerne, und sie lieben gerne. Sie haben Freude am Leben. Und diese Lebenslust drücken sie aus, auch im Design. Ein Design-Genie allerdings ist kein Italiener. Es ist der Franzose Philippe Starck, von dem Anna Castelli-Ferrieri sagte: „Philippe Starck ist ein Künstler." Er hat Dinge verwirklicht, von denen ich immer geträumt habe, sie zu machen.

Sie arbeiteten einige Jahre in der Werbeagentur TBWA in Frankfurt im Bereich Verpackungsdesign. Was machten Sie dort?

Ich lernte mit dem Begriff Marketing umzugehen, denn davon hatte ich keine Ahnung. Sie können heute vieles nicht auf den Markt bringen, wenn das Marketing nicht stimmt. Nehmen Sie zum Beispiel meine Topfserie für Spring, die – obwohl mit Designpreisen ausgezeichnet – ein großer Flop wurde. Die Kochgilde war begeistert von diesen Töpfen, die auf engstem Raum zu stapeln und doppelstöckig zu bekochen sind. Doch das Marketing war schlecht. „Die neue Lust am Kochen", hieß der Werbeslogan. Was sollte daran neu sein? Da muss ich doch der Hausfrau sagen: „Hör zu, Du kaufst nur vier Kochtöpfe, mehr brauchst Du nicht zum Kochen und sparst obendrein noch Platz im Schrank."

In einer Selbstdarstellung betiteln Sie einen Teil ihrer Arbeit als „Architecture Design". Was bedeutet dieser Begriff, und was verbindet Architektur und Design?

Wo fängt Design an und wo hört Architektur auf? Deswegen gibt es das Wort Architekturdesign. Also Dinge, die zur Architektur gehören, wie Türen, Beschläge etc. Auch Messestände und Verkaufsräume zähle ich zu Architekturdesign.

enthusiasm for design would be greater in Germany too. And then the Italians love to eat, and they love loving. They take joy in living. And they express this love of life – also in design. Mind you, one design genius is no Italian. He is the Frenchman Philippe Starck, of whom Anna Castelli-Ferrieri said: "Philippe Starck is an artist." He has created things that I have always dreamt of making.

You worked for some years for the advertising agency TBWA in Frankfurt in the area of packaging design. What did you do there?

I learnt to handle the concept of marketing, as I had had no idea previously. Today you can not bring much onto the market if the marketing is not right. Take for example my ranges of pots and pans for Spring, which – despite having been awarded designer prizes – was a great flop. The cooking guild was enthusiastic about these pans, which stacked to fit in the smallest space and could be used for cooking with one on top of the other. But the marketing was bad. The advertising slogan was "New pleasure in cooking". What was supposed to be new about it? Then I have to say to the housewife: "Listen, you buy just four pans, you don't need more for cooking, and on top of that you save more space in the cupboard."

In a description of yourself, you gave part of your work the title "Architecture Design": What does this concept mean and what connects architecture and design?

Where does design start and architecture stop? That is why there is the expression architectural design. So I mean things which belong to architecture like doors, fittings etc. I also count exhibition stands and showrooms as architectural design. Design is actually everything; it is design of life, like architecture. When Ikea became popular in the early seventies I thought, damn me, why is there no house at Ikea that I can build myself?

Design ist eigentlich alles, ist Gestaltung des Lebens, wie die Architektur. Als Ikea Anfang der siebziger Jahre populär wurde, dachte ich mir, verflucht noch mal, warum gibt's bei Ikea kein Haus, das ich mir selber zusammenbauen kann? Wir haben eine solche Restriktion von bürokratischen Bauvorschriften, die fast jegliche Kreativität in der Architektur erstickt. Und was macht Philippe Starck? Der konstruiert ein Haus, das sich jeder selber bauen und leisten kann. Ich denke, wenn wir über Design reden und letzten Endes auch über Architektur, ist das Wichtigste die Funktion. Das Dach gegen Regen und Schnee, die Heizung gegen Kälte, die Fenster für das Licht und frische Luft. Erst dann beginnt das Gestalten.

Also steht Funktionalität ganz vorne?

Ja, absolut. Aber warum darf ein Produkt nicht auch eine gewisse Sinnlichkeit haben? Gerade die Sinnlichkeit und die Emotionalität eines Produktes sind unglaublich wichtig. Ich behaupte, dass ein Produkt Psyche hat. Meine Großmutter trank zum Beispiel Tee nur aus einer ganz bestimmten silbernen Teekanne. Als ich die Mono-Teekanne entworfen hatte, habe ich sie ihr zum Testen gegeben. Nach vierzehn Tagen sagte sie: „Ich benutze nur noch Deine Teekanne!" Da wusste ich, das Ding ist ein Hit. Daran sehen Sie, dass Emotionen beim Gestaltungsprozess ganz wichtig sind.

Für Lauscha Glaskunst entwarfen Sie ein Champagner-Glas namens „Christiane". Sieht ein bisschen phallisch aus, oder?

Kann man so sehen oder auch nicht. Hier hat die Natur etwas mitgespielt. Wenn die Calla eine phallische Blume ist, dann haben Sie sicherlich Recht. Wichtig ist ja nicht nur das Aussehen, sondern auch die Funktion. Mein Champagnerglas hat zwei Kelche. Aus dem einen trinken Sie Champagner, und – wenn Sie es umdrehen – können Sie daraus zum Beispiel Schnaps trinken. In der Presse wurde es vereinzelt als frauen-

We have such a restriction in bureaucratic building regulations, which stifles almost all creativity in architecture. And what does Philippe Starck do? He constructs a house that anyone can build and afford. When we are talking about design – and also after all about architecture – I think that the most important thing is the function. The roof keeps out rain and snow, the heating keeps out cold, windows are for light and fresh air. Only then can designing begin.

So functionality is absolutely key?

Yes, certainly. But why should a product not also be allowed to have a certain sensuality? In particular the sensuality and emotionality of a product are unbelievably important. I would say that a product has a psyche. For example, my grandmother only drank tea from a quite distinctive silver teapot. When I had designed the Mono teapot, I gave it to her to test. After two weeks she said: "I'm only using your teapot." Then I knew that the thing is a hit. So you see that emotions are most important in the design process.

You designed a champagne glass with the name "Christiane" for Lauscha Glaskunst. Looks a bit phallic, doesn't it?

You can look at it that way or not. Nature has had a part to play. If the calla is a phallic flower, you are certainly right. It is not just what it looks like that is important but also the function. My champagne glass has two goblets. From one you drink champagne, and if you turn it round you can drink schnapps out of it, for instance. It was sometimes portrayed as anti-women in the press. I do not know what is supposed to be anti-women about the glass. I mean, sexuality has something to do with desire and sensuality of life. Take the gas lighter by Alessi – also a kind of phallic symbol.

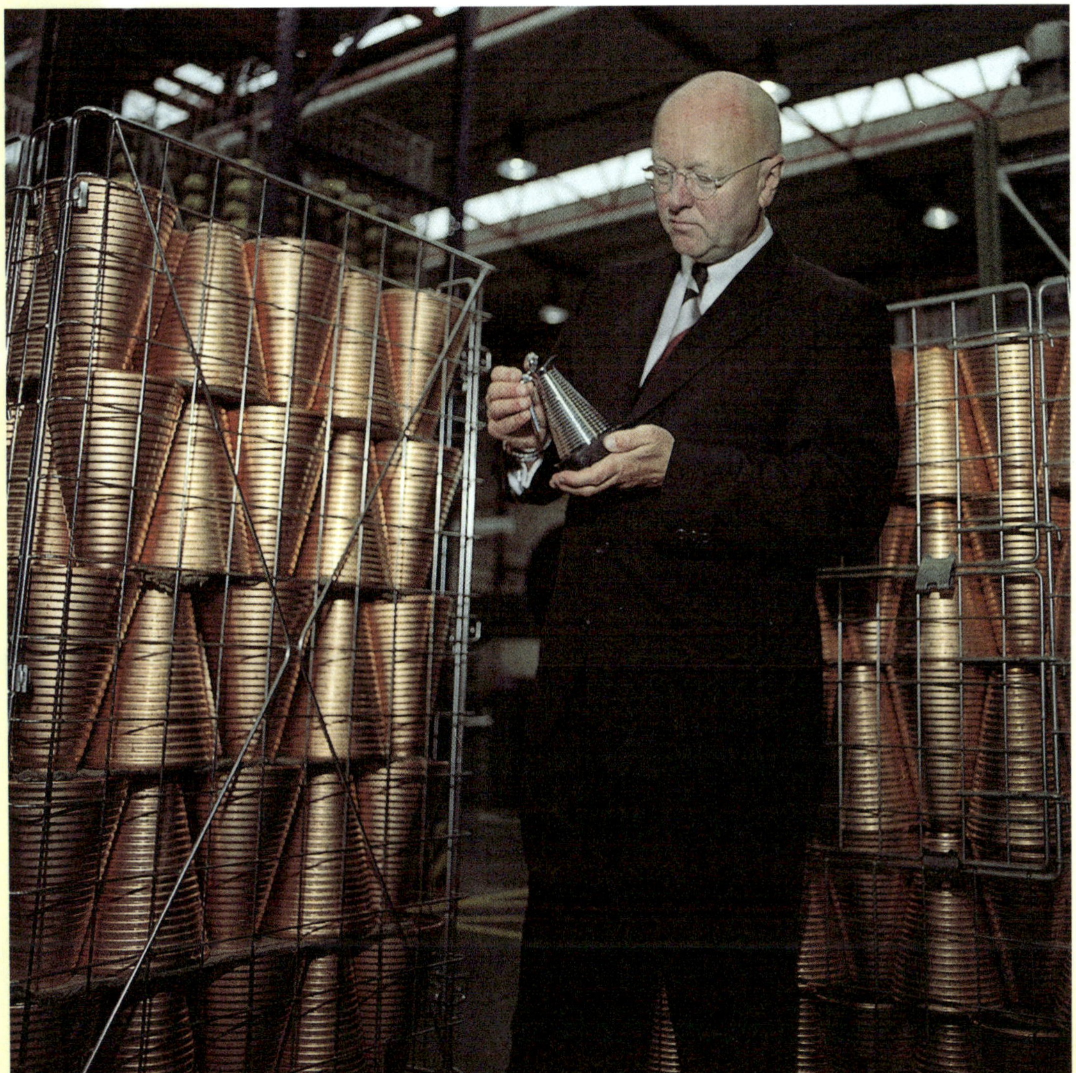

Jörg Dümmig-Zitzmann
Geschäftsführer Alfi-Zitzmann GmbH
Managing director Alfi-Zitzmann GmbH

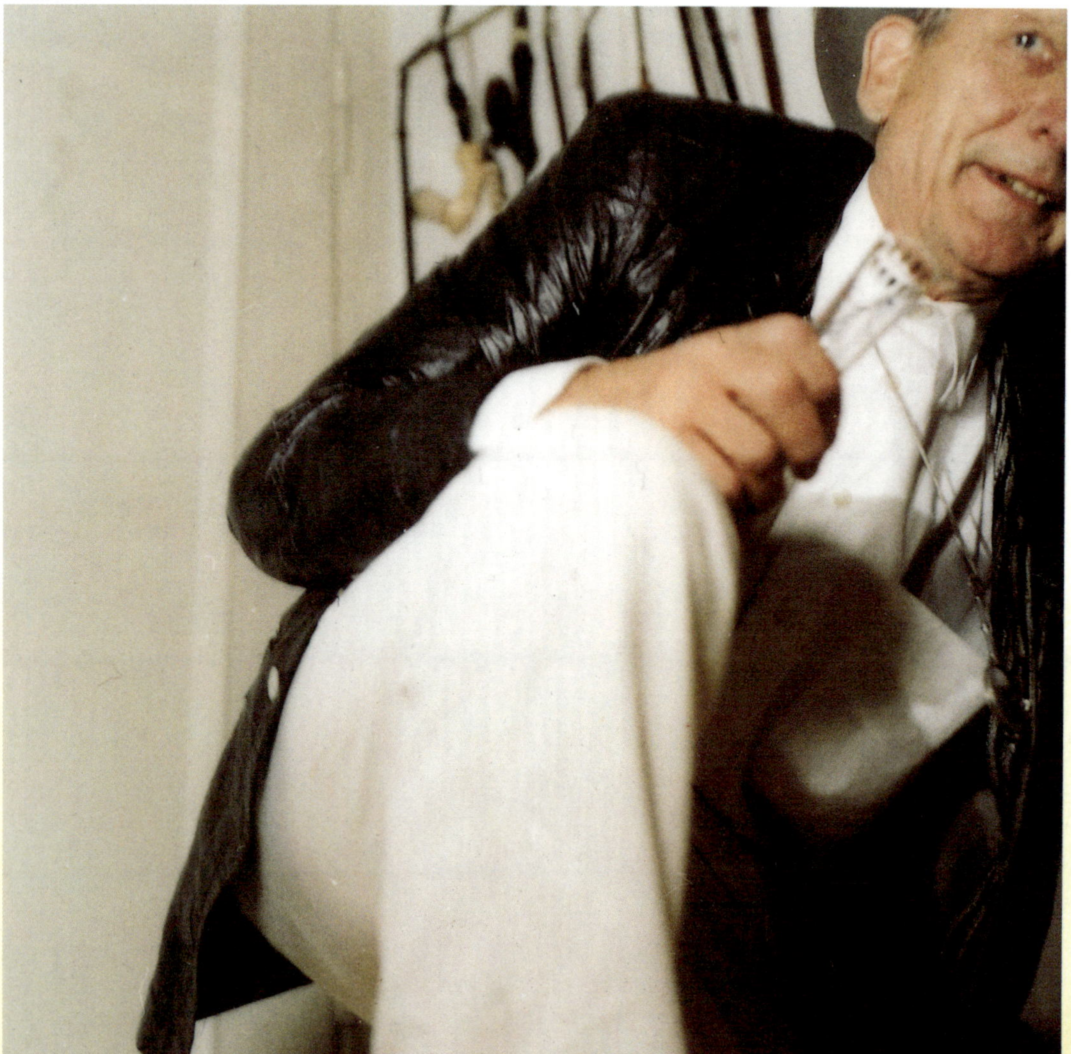

Prof. Kurt Weidemann
Unternehmensberater
Managing consultant

feindlich bezeichnet. Ich weiß nicht, was an dem Glas frauenfeindlich sein soll. Ich meine, die Sexualität hat ja auch etwas mit Lust und Sinnlichkeit des Lebens zu tun. Nehmen Sie den Gasanzünder von Alessi – auch so eine Art Phallus-Symbol.

1990 gründeten Sie den Deutschen Designer Club. Was war Ihre Intention dabei?
Ich wollte nicht noch einen Verein wie den Verband Deutscher Industriedesigner oder den Deutschen Designer Verband mit fürchterlichen, endlosen Sitzungen und Hauptversammlungen gründen. Sondern einen Club, in dem sich professionelle Designer der verschiedenen Designdisziplinen finden und gemeinsam die Qualität ihrer Arbeit der Industrie gegenüber bewusst machen. Kontrovers zu diskutieren ist ja schön und gut, aber schlimm ist der Neid unter vielen Designern. Vielleicht, weil sich laut Statistischem Bundesamt in Wiesbaden 90 Prozent unserer Berufssparte am Existenzminimum bewegen. Deswegen frage ich mich, woher der Run auf diesen Berufszweig kommt. Wir müssen Neid und Missgunst abbauen und uns mehr als Familie sehen, dann können wir uns auch der Industrie gegenüber besser verkaufen. Es ist doch nie einer hergegangen und hat gesagt: „Du, hör mal zu, deine Arbeit ist aber genial." Im Gegenteil.

Der DDC als Pendant zum ADC ?
Ich bin seit vielen Jahren Mitglied beim Art Directors Club, und das hat mich natürlich inspiriert.

Welcher Ihrer ältesten Entwürfe wird immer noch produziert?
„Bonanza" von Berndes zum Beispiel. Gehen Sie mal in den Kaufhof oder zu Lorey in die Topfabteilung und fragen nach der „Bonanza"-Serie. Die ist schon 25 Jahre alt. Das ist toll!

Vielen Dank für dieses kurzweilige Gespräch, Herr von Grolman.
Also ich könnte noch stundenlang weiterreden ...

You founded the German Designer Club (DDC) in 1990. What was your intention there?
I did not want to found just another club like the Verband Deutscher Industriedesigner (Assocation of German Industrial Designers) or the Deutscher Designer Verband (Association of German Designers) with dreadful, endless meetings and AGMs. Rather it was to be a club in which professional designers from the various design disciplines are brought together and in which they can together make industry aware of the quality of their work. Controversial discussions are all well and good, but the jealousy amongst many designers is bad. Perhaps because – according to the German government statistical office in Wiesbaden – 90 percent of our profession barely earn enough to scrape a living. This is why I wonder where the great popularity of this career path is coming from. We have to break down envy and resentment and regard ourselves more as a family, then we can sell ourselves better to industry. It is not as if anybody has ever come up to me and said: "You, listen, your work is just brilliant." Quite the opposite.

Do you see the DDC as a counterpart to ADC ?
I have been a member of the Art Directors Club for many years and of course that inspired me.

Which of your oldest designs are still produced?
"Bonanza" by Berndes for example. Go into a Kaufhof store or to Lorey in the pots and pans department and ask for the "Bonanza" range. That is 25 years old now. That is great!

Thank you for the entertaining conversation, Herr von Grolman.
I could carry on talking for hours ...

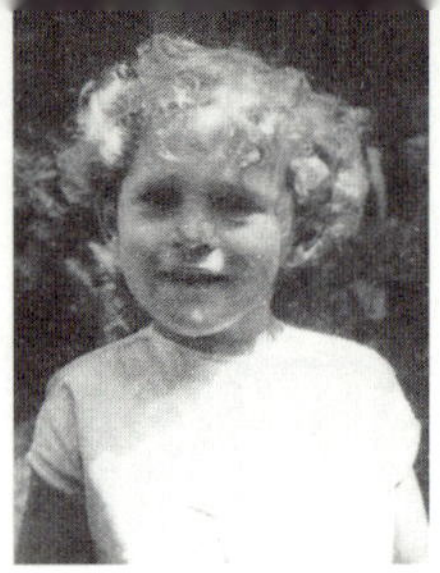
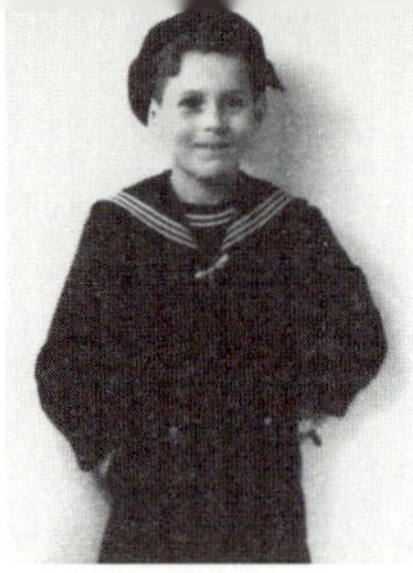
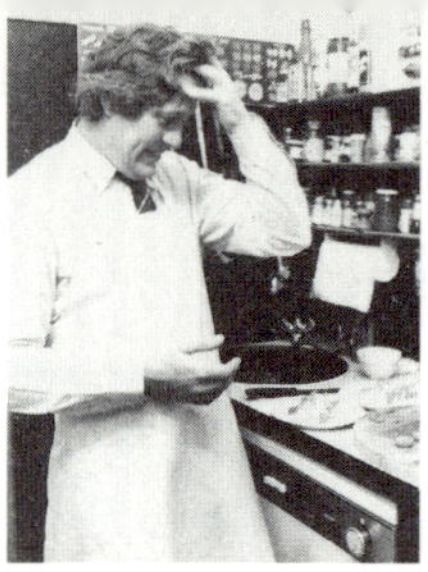
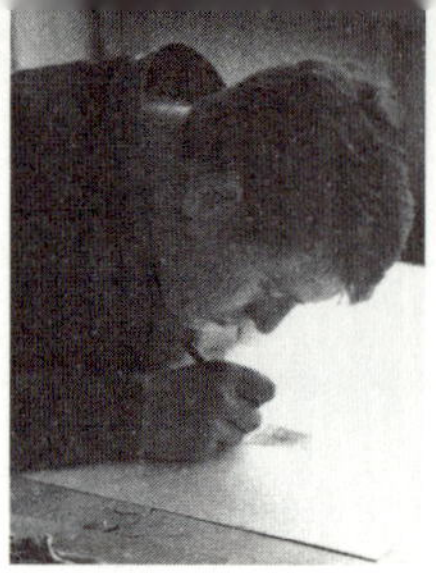
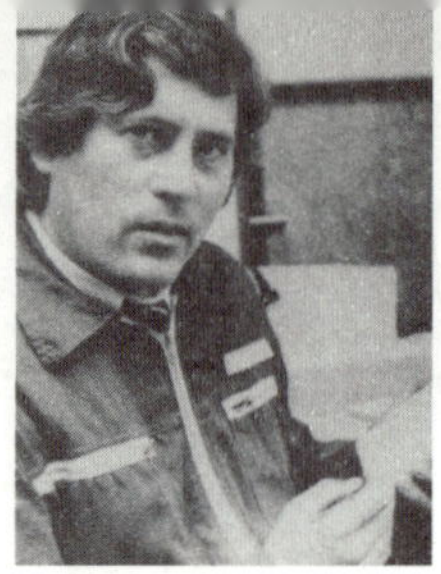
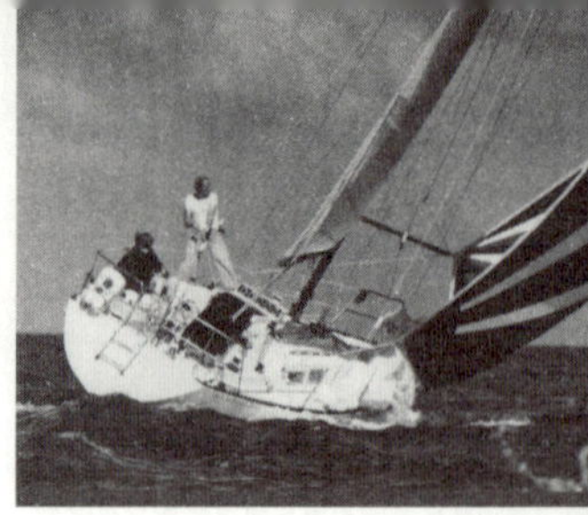
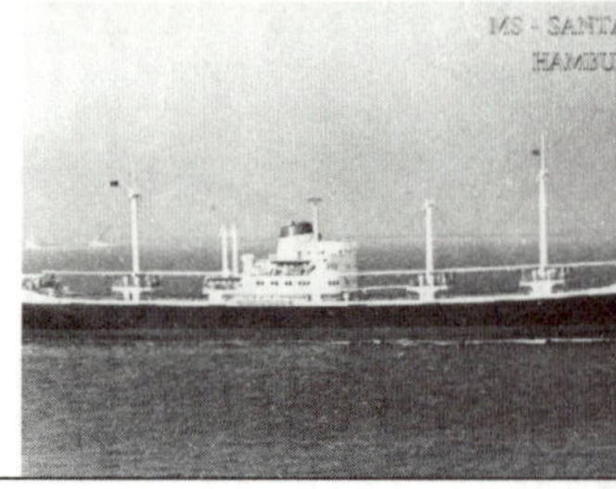

BIOGRAPHIE
BIOGRAPHY

1942
Tassilo von Grolman wird am 23. Dezember in Iserlohn geboren. Seine Schulzeit verbringt er in Detmold, Bonn und Oldenburg.

1959–1962
Ausbildung zum Maschinenschlosser bei der AEG in Oldenburg. Erste kleinere Designarbeiten und konkretes Berufsziel, Designer zu werden. Beginn Maschinenbau-Studium am Technikum in Lage/Lippe (1962). In den Semesterferien Praktika bei Firma Otto Krause in Iserlohn.

1965–1966
Unterbrechung des Studiums und für ungefähr ein Jahr Tätigkeit als Ingenieur-Assistent bei der Reederei Hamburg-Süd. Ausgangsort war Bremen mit Stationen in New York, Sydney, Melbourne, Auckland, Los Angeles, San Francisco, Vancouver, Mexiko, der Südsee und der Karibik. Anschließend Fortführung des Maschinenbau-Studiums an der Staatlichen Technikerschule in Berlin.

1967–1971
Abschluss des Maschinenbaustudiums. Beginn des Industriedesign-Studiums an der Staatlichen Akademie für Werkkunst und Mode in Berlin. Ab 1968 Fortführung des Studiums an der Werkkunstschule, heute Gesamthochschule, in Kassel. Assistent von Prof. Arnold Bode im Bereich Ausstellungsplanung (ab 1970).

1972–1973
Abschluss des Studiums als Industrial Designer. Tätigkeit im Architekturbüro Wolfgang Bader in Frankfurt/Main. Freelance-Tätigkeiten bei den Werbeagenturen TBWA und Lürzer, Conrad in Frankfurt am Main. Heirat mit Cam-Hong Bui-Thi.

Seit 1974
Auslandsaufenthalte in Tokio/Japan und partnerschaftliche Zusammenarbeit mit Kichinosuke Shikuwa, Norm-Design.

1975
Gründung des eigenen Designbüros in Frankfurt/Main.

Seit 1979
Partnerschaftliche Zusammenarbeit mit Peter Engelhardt, Ideal-Design.

1980
Designentwicklung der Teekanne mit Siebeinsatz.

1981
Heirat mit Dagmar Maßmann. Gemeinsamer weiterer Aufbau des Designbüros. Geburt von Alexandra Nora von Grolman und Umzug des Designbüros nach Oberursel bei Frankfurt/Main.

1984
Auszeichnung für hohe Designqualität des Design Zentrums Nordrhein Westfalen für den Haartrockner „Allidry Studio W 022“. Jurymitglied des Fachverbandes Kunststoff Konsumwaren (bis 1997).

1985
Vorstandsmitglied beim Deutschen Werkbund Hessen (bis 1996). Geburt von Victoria Philippa von Grolman.

1986–1987
Aufnahme der Teekanne „Mono Lab 2“ in die Ständige Sammlung des Museums für Kunst und Gewerbe Köln. Aufnahme der Teekanne „Mono Tea 450“ in die 20th Century Design and Architecture Collection des Metropolitan Museum of Art, New York.

1988
Gastdozent für Verpackungsdesign an der Fachhochschule Darmstadt.

Seit 1989
Jurymitglied des Art Directors Club von Deutschland.

1990–1997
Gründung des Deutschen Designer Clubs (DDC) mit Sitz in Oberursel und Übernahme der Präsidentschaft (bis 1997).

1991
Gastdozent an der Hochschule für Bildende Künste in Lissabon. Jurymitglied „Design Arena Leuchten“ der Zeitschrift „Zuhause“. Teekanne „Mono Tea 450“ Exponat in der Ausstellung „Design ... made in Germany“ des Korea Exhibition Center in Seoul, Südkorea.

1993–1994
Teekanne „Mono Filio“ Exponat in der Ausstellung „Vierzig Jahre deutsche Designkultur“ des Rat für Formgebung in Frankfurt/Main. Koch- und Serviergeschirr „Compact GLI“ ausgezeichnet als „Produkt des Jahres“ von der „Kookgilde“, Niederlande.

1995
Ausstellung „Design und Innovation – Tassilo von Grolman“ in der Securitas-Galerie in Bremen. Good Design Award des The Chicago Athenaeum für die Isolierkannen „Big Mama“ und „Hotel-Design“ und für das Koch- und Serviergeschirr „Compact GLI“, Silbermedaille des Art Directors Club für das Packungs- und Formdesign mit Broschüre der „Twinings-Tea-Caddy“-Künstler Edition.

1999
Gastdozent für Designgeschichte an der Akademie der Bildenden Künste in Nürnberg. Good Design Award des The Chicago Athenaeum für den Zigarrenaschenbecher „Sancho“.

1984–2000
Zahlreiche nationale und internationale Design-Auszeichnungen, Ausstellungen, Design-Wettbewerbe und Produktauswahlen. Mitgliedschaften im Deutschen Werkbund Hessen, Art Directors Club von Deutschland, Design-Zentrum Hessen.

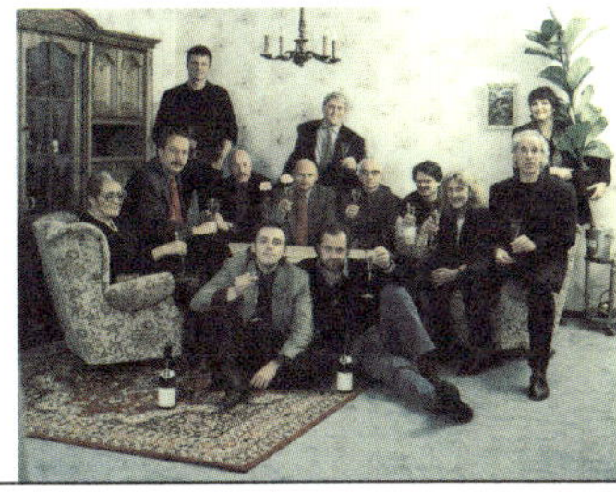 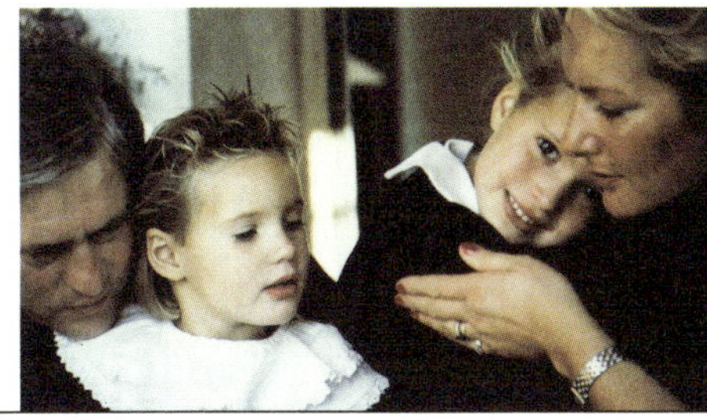

1942
Tassilo von Grolman is born on the 23rd of December in Iserlohn. He spends his schooldays in Detmold, Bonn and Oldenburg.

1959–1962
Training as an engine fitter with AEG in Oldenburg. First small design projects and concrete career decision to become a designer. Beginning of engineering studies at the Technikum in Lage/Lippe (1962). Practical experience during the holidays with the Otto Krause Company in Iserlohn.

1965–1966
Break from studies, work as an engineer's assistant with the shipping company Hamburg-Süd for approximately one year. Starting point was Bremen with stops in New York, Sydney, Melbourne, Auckland, Los Angeles, San Francisco, Vancouver, Mexico, the South Pacific and the Caribbean. Afterwards continuation of engineering studies at the Staatliche Technikerschule in Berlin.

1967–1971
Completion of engineering studies. Beginning of study of industrial design at the Staatliche Akademie für Werkkunst und Mode in Berlin. From 1968 continuation of studies at the Werkkunstschule, now Gesamthochschule, in Kassel. Assistant to Prof. Arnold Bode in the area of exhibition planning (from 1970).

1972–1973
Completion of studies as an industrial designer. Work in the architect's office Wolfgang Bader in Frankfurt/Main. Freelance work with the advertising agency TBWA and Lürzer, Conrad in Frankfurt/Main. Marriage to Cam-Hong Bui-Thi.

Since 1974
Spends time overseas in Tokyo/Japan and works in partnership with Kichinosuke Shikuwa, Norm-Design, Tokyo.

1975
Formation of own design studio in Frankfurt/Main.

Since 1979
Works in partnership with Peter Engelhardt, Ideal-Design.

1980
Development of design of teapot with strainer insert.

1981
Marriage to Dagmar Maßmann. Together see further growth of the design company. Birth of Alexandra Nora von Grolman and relocation of the design company to Oberursel near Frankfurt/Main.

1984
Award for high design quality from the Design Zentrum Nordrhein Westfalen for the hairdryer "Allidry Studio W 022". Member of the jury panel of the Fachverband Kunststoff Konsumwaren until 1997.

1985
Board member of the Deutscher Werkbund Hessen until 1996. Birth of Victoria Philippa von Grolman.

1986–1987
The teapot "Mono Lab 2" becomes part of the permanent collection of the Museum für Kunst und Gewerbe Köln. The teapot "Mono Tea 450" becomes part of the 20th Century Design and Architecture Collection of the Metropolitan Museum of Art, New York.

1988
Visiting lecturer in packaging design at the Fachhochschule Darmstadt.

Since 1989
Member of the jury panel of the Art Directors Club of Germany.

1990–1997
Formation of the Deutscher Designer Club with office in Oberursel, and acceptance of the presidency (until 1997).

1991
Guest lecturer at the Hochschule für Bildende Künste in Lisbon. Member of the judging panel for the "Design Arena Leuchten" of the magazine "Zuhause". Teapot "Mono Tea 450" exhibit at the exhibition "Design ... made in Germany" of the Korea Exhibition Centre in Seoul, South Korea.

1993–1994
Teapot "Mono Filio" exhibit at the exhibition "forty years of German design culture" of the Rat für Formgebung in Frankfurt. Cooking and serving set "Compact GLI" wins title of "Product of the Year" of the "Kookgilde", Netherlands.

1995
Exhibition "Design and Innovation – Tassilo von Grolman" in the Securitas-Galerie in Bremen. Good Design Award from The Chicago Athenaeum for the thermos flask Big Mama" and "Hotel-Design" and for the cooking and serving set "Compact GLI", silver medal of the Art Directors Club for packaging and design for brochure of the "Twinings-Tea-Caddy" artists edition.

1999
Guest lecturer in history of design at Akademie der Bildenden Künste in Nuremberg. Good Design Award from The Chicago Athenaeum for the cigar ashtray "Sancho".

1984–2000
Many national and international design awards, exhibitions, design competitions and product selections. Membership of the Deutscher Werkbund Hessen, Art Directors Club of Germany, Design-Zentrum Hessen.

tassilo v. grolman

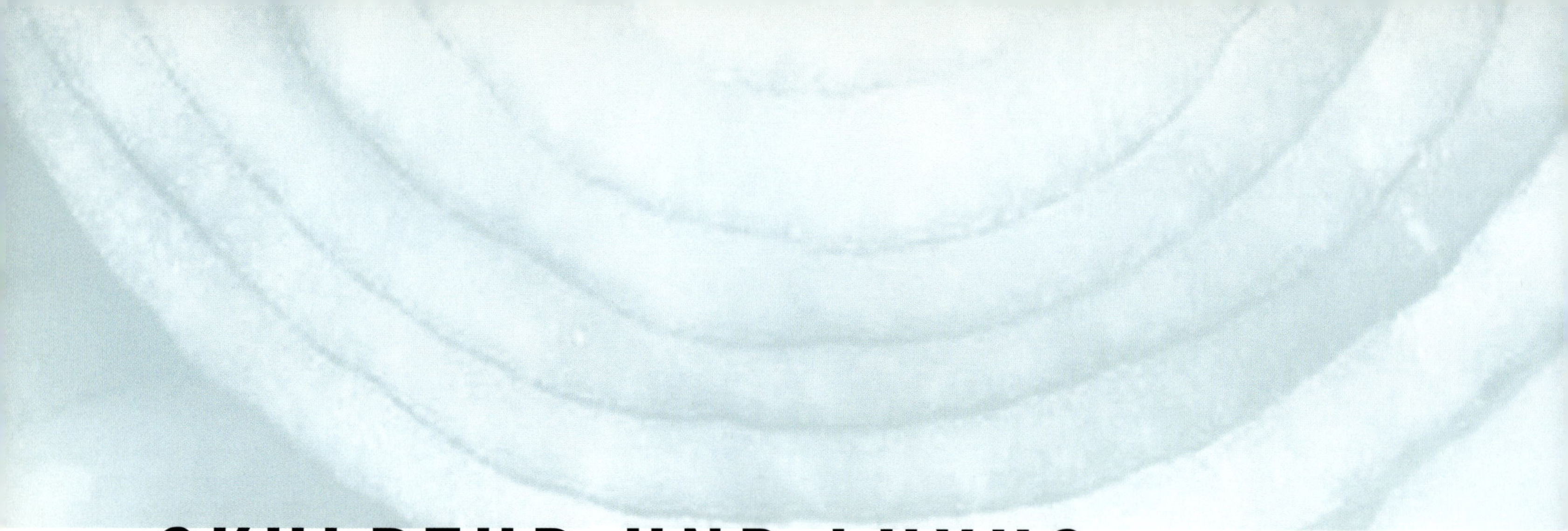

SKULPTUR UND LUXUS

LACHS-TATAR MIT KARTOFFEL-GRATIN

400 g frisches Lachsfilet
100 g Keta-Lachskaviar
1,5 kg Kartoffeln
150 g Butter
50 ml Crème fraîche
0,7 l Milch
Fleischfond oder
3 Fleischbrühwürfel
6 Knoblauchzehen
frischer Rosmarin
1 kleiner Bund Dill
4 Frühlingszwiebeln
1 mittelgroße Zwiebel
1 Bund Schnittlauch
1/2 Zitrone
Salz
Pfeffer

1 Den Backofen auf 180 °C vorheizen, Gas Stufe 2. Die Kartoffeln schälen und mit einem Gurken- oder Trüffelhobel in Scheiben schneiden. Den Knoblauch schälen und klein hacken.

2 Eine Kasserolle mit einem Teil der Butter leicht einfetten. Crème fraîche, Milch, Knoblauch, Rosmarin und Fleischfond gut verrühren, die Kartoffeln unterheben und in die Kasserolle füllen.

3 Die Kartoffeln mit Butterflöckchen bedecken und ungefähr 1 1/4 Stunden garen. Den Dill waschen.

4 Die Zwiebel und die Frühlingszwiebeln schälen und die Zwiebel sehr fein hacken. Den Lachs enthäuten und entgräten, in kleine Streifen schneiden und mit einem großen Küchenmesser sehr fein hacken – ähnlich wie man Petersilie klein hackt – so dass Tatar entsteht. Kein zusätzliches Öl verwenden, weil das Öl des Lachses das Tatar zusammenhält und noch dazu das gesündeste überhaupt ist!

5 Mit Salz und Pfeffer abschmecken. Die Masse zu einem Rund formen, Zitronensaft darüber träufeln und mit Dill und Keta-Lachskaviar dekorieren. Außenherum einen Kranz aus den gehackten Zwiebeln bilden und als Abschluss diesen mit den Frühlingszwiebeln garnieren.

6 Das Tatar zum Kartoffel-Gratin servieren. Dazu passt immer ein erfrischender Chablis, Riesling oder Pinot Grigio.

RENATE PUVOGEL

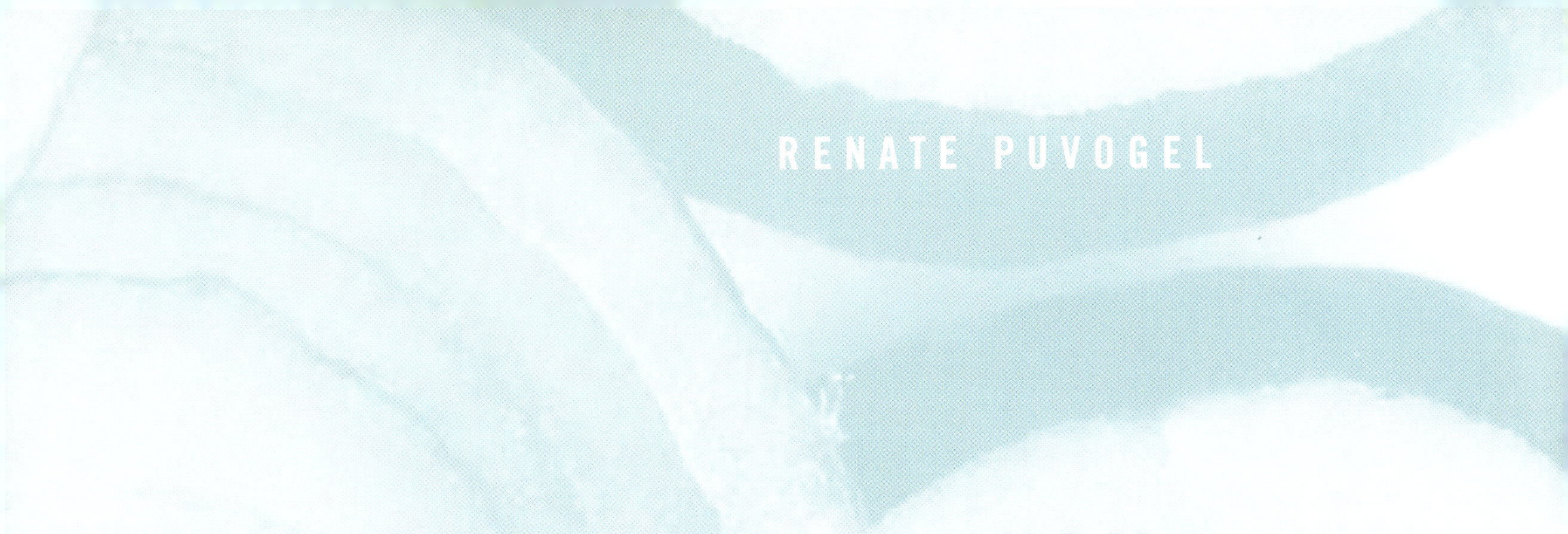

SCULPTURE AND LUXURY

- 400 g fresh salmon fillet
- 100 g Keta-salmon caviar
- 1.5 kg potatoes
- 150 g butter
- 50 ml crème fraîche
- 0.7 l milk
- Meat stock or
- 3 meat stock cubes
- 6 cloves of garlic
- Fresh rosemary
- 1 small bunch of dill
- 4 spring onions
- 1 medium sized onion
- 1 bunch chives
- 1/2 lemon
- Salt
- Pepper

SALMON TARTARE WITH POTATO GRATIN

1 *Pre-heat the oven to 180°C, gas mark 2. Peel the potatoes and slice with a cucumber or truffle slicer. Peel the garlic and chop finely.*

2 *Lightly grease a casserole dish with part of the butter. Mix the crème fraîche, milk, garlic, rosemary and meat stock well together, fold in the potatoes and pour into the casserole dish.*

3 *Cover the potatoes with small knobs of butter and bake for around 1 1/4 hours. Wash the dill.*

4 *Peel the onions and spring onions and chop the onions very finely. Skin and bone the salmon, cut into small strips and chop very finely using a large kitchen knife – in the same way that you would finely chop parsley – so that it turns into tartare. Do not use additional oil because the oil from the salmon holds the tartare together – and is also the healthiest part of all!*

5 *Season with salt and pepper. Form the mixture into a round shape, trickle some lemon juice over it and decorate with dill and Keta-salmon caviar. Make a wreath around the edge from the chopped onions and finally garnish with spring onions.*

6 *Serve the tartare with the potato gratin. A refreshing Chablis, Riesling or Pinot Grigio goes well with this dish.*

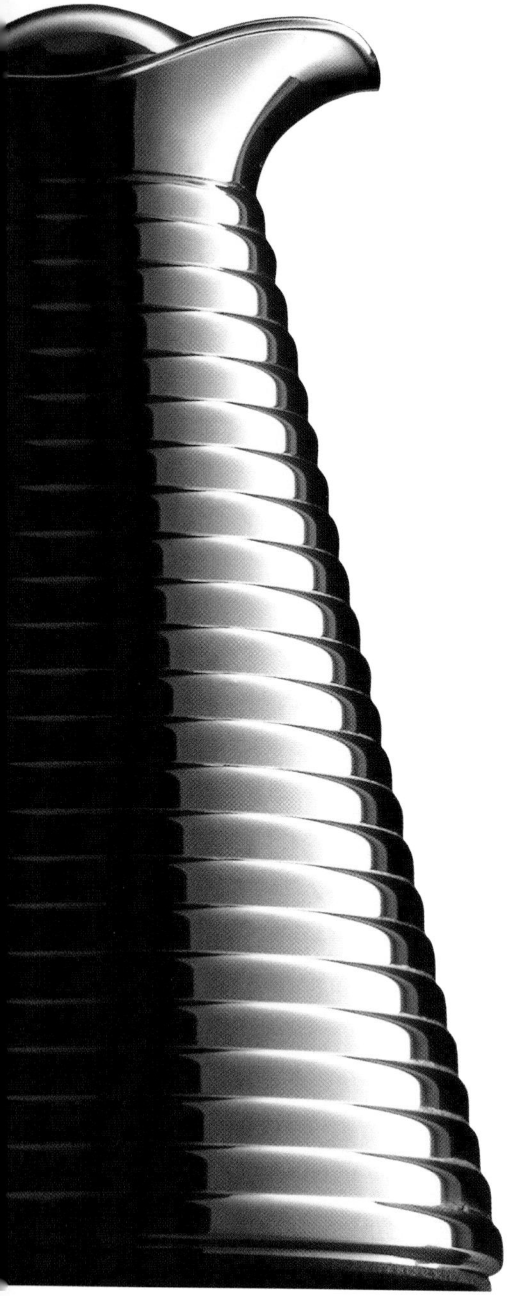

Linke Seite
Isolierkanne „Achat" für Alfi, 1990

Rechte Seite
Isolierkanne „Big Mama" für Alfi, 1993.

Left-hand side
Thermal pot "Achat" for Alfi, 1990.

Right-hand side
Thermal pot "Big Mama" for Alfi, 1993.

DESIGNER-KÜNSTLER UND KÜNSTLERISCHES DESIGN

Wie einst in den zwanziger Jahren bewegen sich seit einiger Zeit die Bereiche von angewandter und freier Kunst wieder vielfältig aufeinander zu. Bemerkenswert viele Museen widmen sich innerhalb ihres Ausstellungsprogramms Strömungen, Themen oder Schöpfern aus dem Designbereich. So wurden klingende Namen wie Ettore Sottsass und Philippe Starck nicht nur Insidern des Designs ein Begriff. Aus der Gegenrichtung, der Kunst kommend, engagieren sich Künstler wie der Kalifornier Jorge Pardo oder der Deutsche Tobias Rehberger intensiv auf dem Feld ästhetischer Raumgestaltung. Beide Künstler entwerfen Lampen, Tische und Sofas, schaffen Modelle für besondere Wohnsituationen und laden in rundum ausgestattete Lokalitäten ein. Ein anderer Kunsteleve, Matti Braun, gestaltete, man staune, den runden Tisch und das gesamte Ambiente der G7-Konferenz in Köln. Die Grenzen zwischen Kunst und Design sind somit fließend, und nicht von ungefähr weist gutes Produktdesign Qualitäten einer Plastik auf. Über das Funktionale hinaus sollte ein Objekt also auch ganz es selbst sein, sich ausleben können in seiner ureigenen, niemandem verantwortlichen, anschaulichen Form. Ein schöner Gegenstand ergibt sich durch die Wahl und Kombination angenehmer Materialien, durch die Ausgewogenheit der einzelnen Details. Tassilo von Grolmans Tee- wie auch Kaffeekannen für „Mono Tischkultur" entsprechen diesen Forderungen vollkommen. Vergleicht man etwa seine durch ihre Riffelung vielfältig reflektierenden Isolierkannen „Achat" und „Big Mama" miteinander, so hat ihr Designer mit „Achat" eine majestätisch aufragende, männliche Kanne geschaffen, die einen geraden, parallel zu dem konisch sich verjüngenden Körper verlaufenden Griff hat; das weiblich gerundete Pendant „Big Mama" hingegen ist mit einem Henkel ausgestattet, der die Kurvatur in Gegenrichtung aufgreift und in seiner Außenbegrenzung – passend zum Partner – wiederum die Gerade aufnimmt: Die Partner passen zueinander und sind dennoch in Spannung gehalten.

DESIGNER-ARTISTS AND ARTISTIC DESIGN

Just as once was the case in the twenties, the areas of applied and free art have been variously moving closer to each other for some time. Many museums notably dedicate space within their exhibition program to trends, themes or creators from the area of design. In this way, names such as Ettore Sottsass and Philippe Stark became familiar even to those outside the circle of design-insiders. Coming from the other side, from art, artists such as the Californian Jorge Pardo or the German Tobias Rehberger are heavily involved in the field of aesthetic interior design. Both artists design lamps, tables and sofas, create objects for particular living areas and invite people into complete fitted out facilities. Another student of art, Matti Braun, designed – astoundingly – the round table and entire ambience of the G7 conference in Cologne. The borders between art and design are thus fluid, and it is no accident that good product design contains sculptural qualities. Besides being functional, an object should also be itself, be able to come to life in the expression of its own, vivid, responsible-to-no-one design. A beautiful object is the result of the choice and combination of suitable materials, and the balance of the individual details. Tassilo von Grolman's teapots and his coffeepots for the "Mono table culture" meet these requirements perfectly. Comparison, for instance, of the thermal pots "Achat" and "Big Mama", whose ripples cause them to reflect in varied ways, determines that in "Achat" the designer created a majestically soaring, masculine pot with a straight handle running parallel to the conical tapering body; and, on the other hand, its counterpart, the feminine, rounded "Big Mama", is fitted with a handle which curves out in the opposite direction and – fitting with its partner – picks up the straight again on its outer side. The partners fit together but are held in tension.

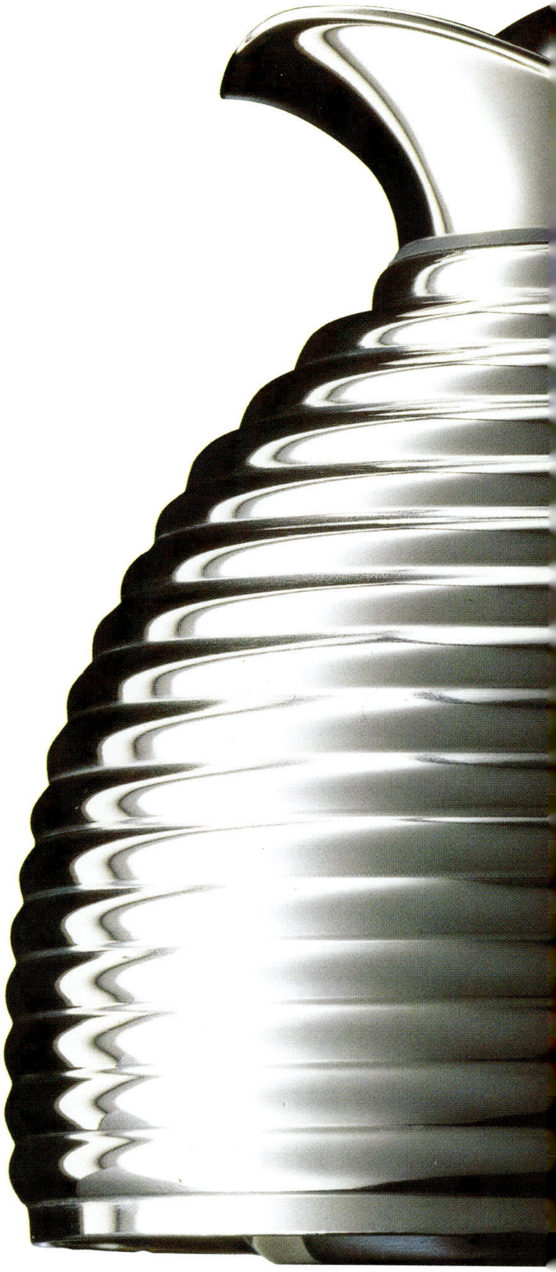

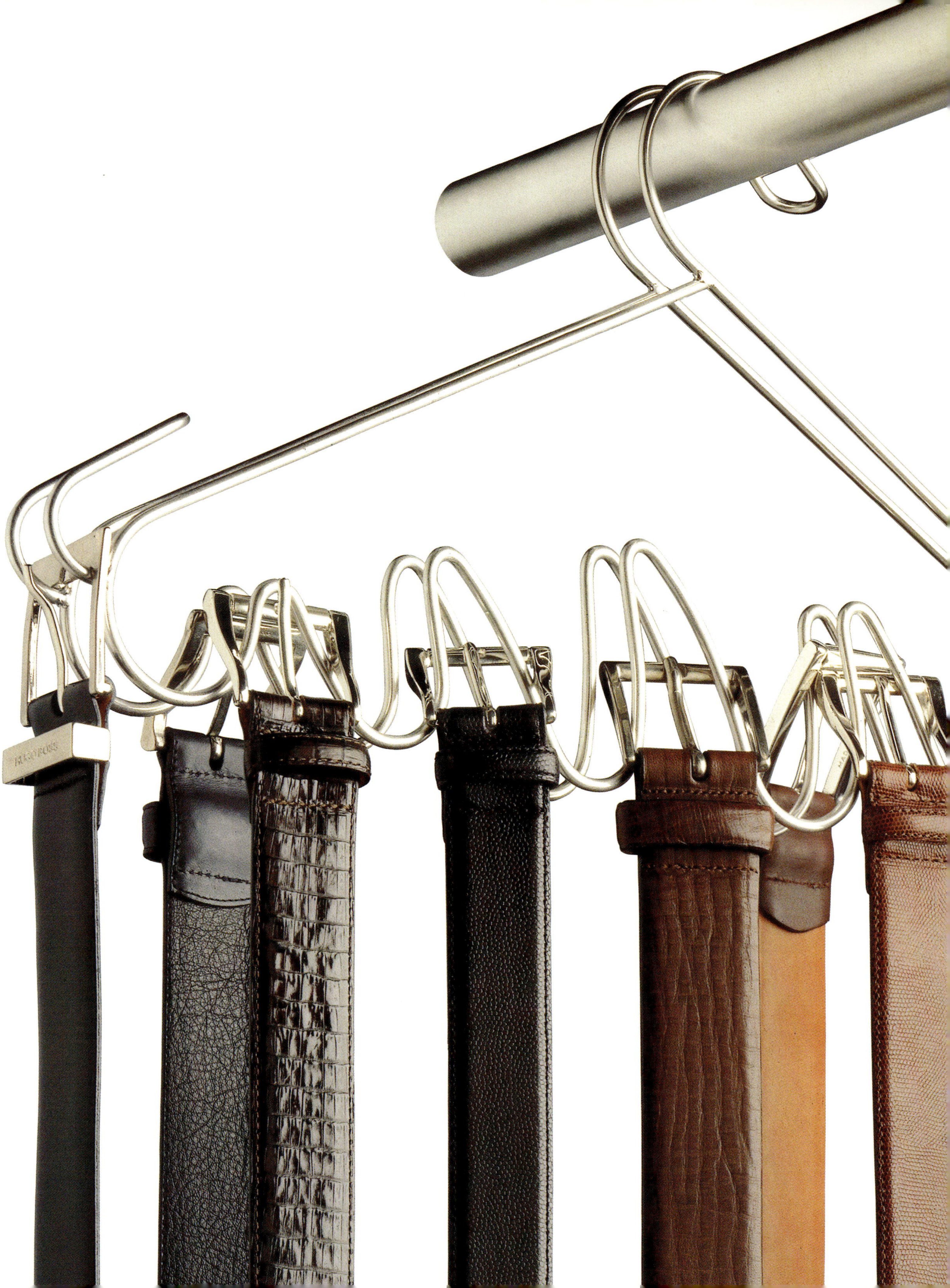

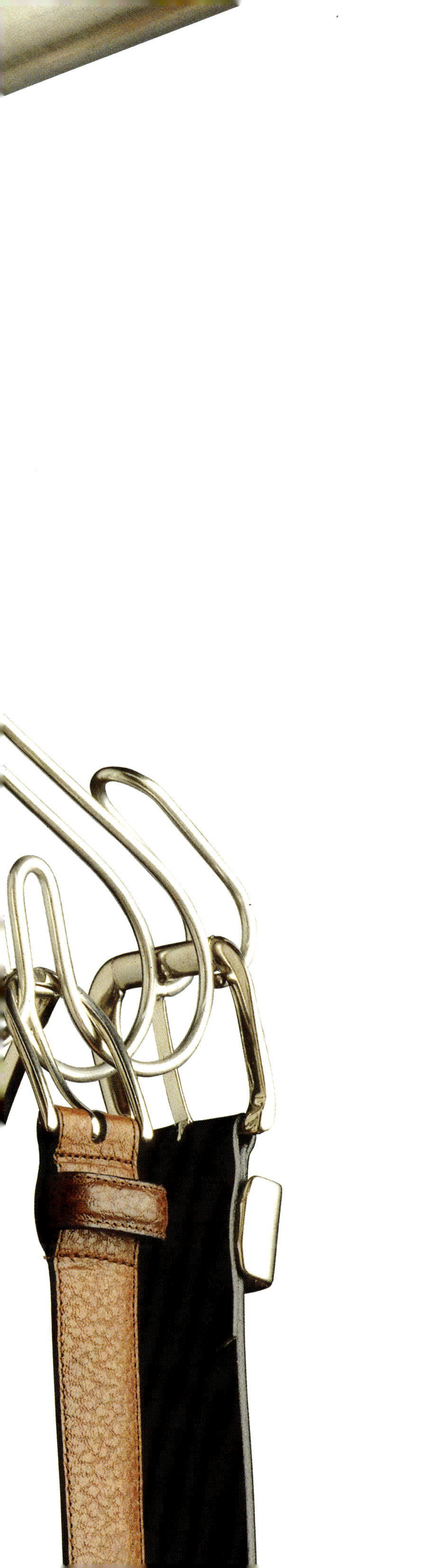

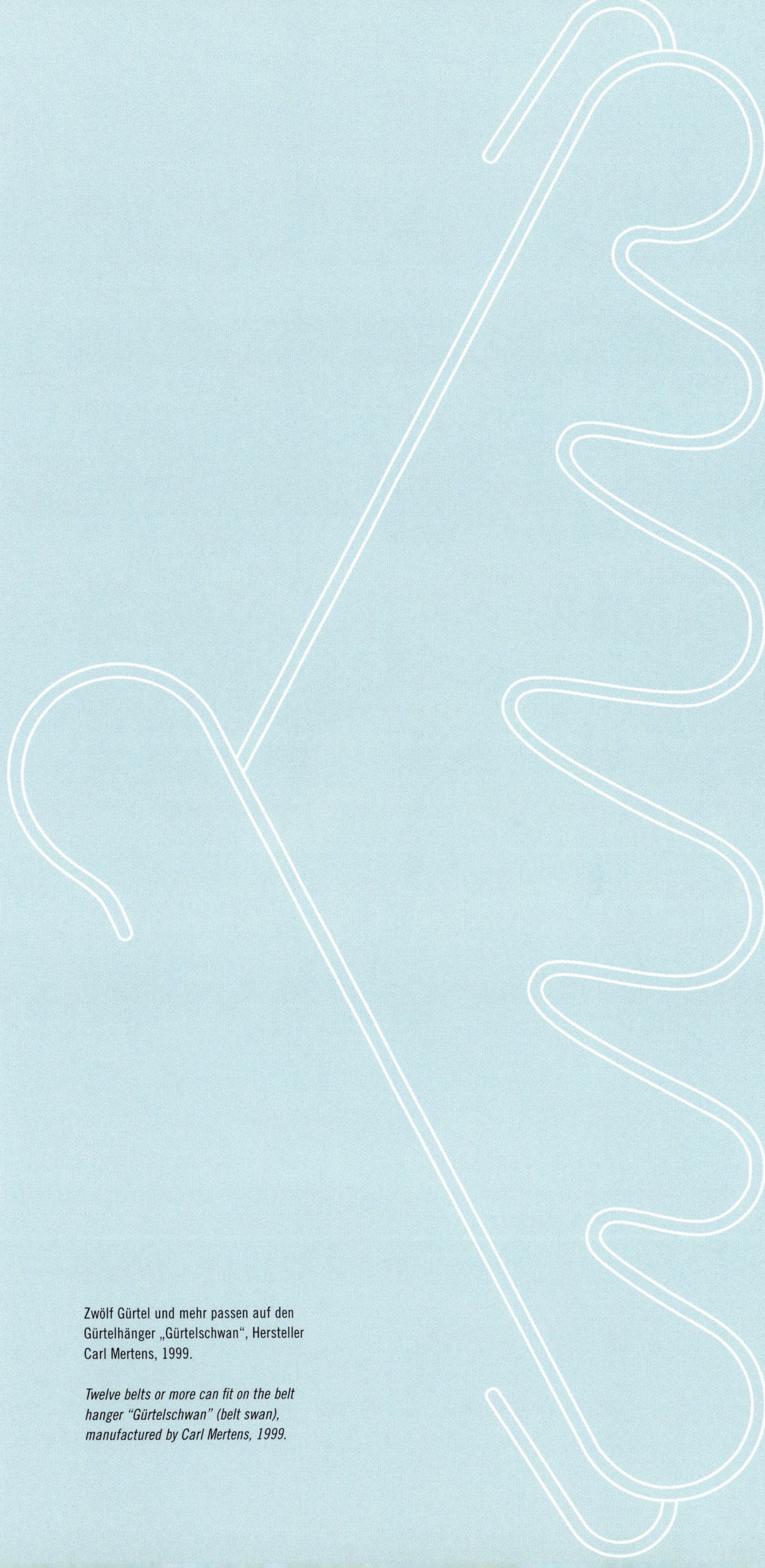

Zwölf Gürtel und mehr passen auf den Gürtelhänger „Gürtelschwan“, Hersteller Carl Mertens, 1999.

Twelve belts or more can fit on the belt hanger “Gürtelschwan” (belt swan), manufactured by Carl Mertens, 1999.

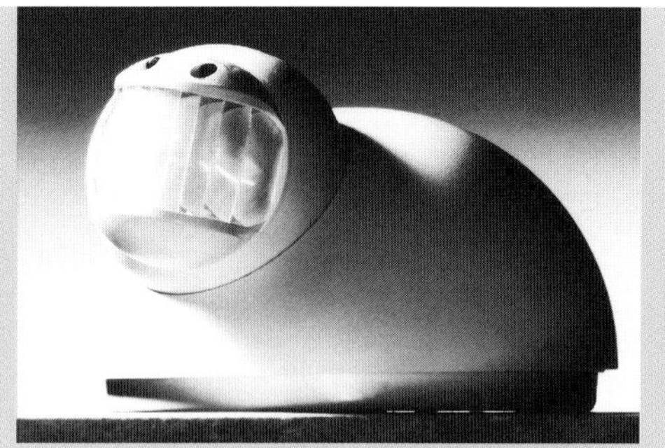

Bewegungsmelder „Mouse Watch", 1984, Hersteller Heinrich Kopp GmbH.

Motion Sensor "Mouse Watch", 1984. Manufactured by Heinrich Kopp GmbH.

FORMENSPIELE UND SPIEL MIT NAMEN

Derlei ästhetische Grundregeln sollte jeder Designschüler idealerweise mit der Muttermilch aufgesogen haben. Ein erfahrener Designer wie Tassilo von Grolman kennt solche Prinzipien nicht nur aus dem Effeff, er liebt es, sie gelegentlich auszureizen und – ohne sie etwa auszuhebeln – eine vorgefundene Form mit einem freien Spiel der Linien aufzufüllen. Seine Gabe, sämtliche Sinne auszukosten, seine eigene Genussfreudigkeit, sein Faible für Eleganz und Opulenz lassen ihn oft genug die nüchterne Konstruktion alltäglicher Gegenstände verlassen und sich Dingen zuwenden, die beileibe nicht lebensnotwendig sind, jedoch das irdische Dasein ungemein bereichern und erquicklicher gestalten. Allein die Parade einiger Entwürfe wie Zigarrenascher, Champagnerkühler oder Récamiere verraten den persönlichen Lebensstil des Designers, der immer der Pflicht gehorcht und gleichzeitig die Kür verehrt, der neben geregelter Alltagsarbeit auch dem Genuss an Augen-, Nasen- und Gaumenfreuden Raum gibt.

Es wäre zum Beispiel verwunderlich, wenn der sporadische Zigarrenraucher von Grolman nur Zigarettenrauchern ein nützliches Utensil erdacht und nicht auch für sich und seinesgleichen alle denkbare Bequemlichkeit für das Niederlegen der Zigarre und das Abstreifen der Asche ersonnen hätte. Der diesbezügliche Designbeitrag ist die materialisierte Bewunderung für einen inzwischen legendären Messing-Entwurf der Bauhaus-Künstlerin Marianne Brandt. Von Grolman stellte dem Brandtschen Entwurf aus dem Jahre 1924 eine Variante aus Silberblech zur Seite. Seine mit wenigen Exemplaren aufgelegte „Hommage à Marianne Brandt" behielt das Prinzip des ursprünglichen „Aschenbechers mit Deckel", nämlich den halb verdeckten Kugeleinsatz, bei und integrierte es in eine langgezogene Form. Der Gegensatz von kreisrundem Aschebehälter und handlichem Griff für den Rauchenden ist aufgehoben. So bettete von Grolman die fülligere Verwandte der Zigarette in eine bequeme Liegestatt. Bei einem zweiten Zigarrenascher, dem Industrieprodukt „Sancho", hat von Grolman genüsslich das Motiv der Rundung im angrenzenden Teil wiederholt und es sogar noch ausgeweitet. Nicht von ungefähr trägt der Ascher aus Aluminium-Sandguss seinen Namen, steht er doch seinem Benutzer stets zu Diensten, ganz so wie seinerzeit sein Namensgeber, der rundliche, ebenso treue wie pfiffige Bauernknappe Sancho Pansa. Der hatte einst seinem Herrn, dem hageren Ritter Don Quijote in Cervantes grandiosem Roman in demütiger Ergebenheit durch sämtliche Abenteuer geholfen. Wie bei „Sancho" klingen oft auch in anderen Produktbezeichnungen der letztlich profanen Gegenstände die beiden Seiten von Grolmans an, nämlich sein praktischer Lebenssinn ebenso wie phantasiebegabter Idealismus. Die erfinderischen, beredten Titel tragen so vielfach zur eindringlichen Wirkung der Objekte ihres Erfinders bei. Sie geben Assoziationen Nahrung, indem sie die in den Dingen schlummernden Ähnlichkeiten mit Figurativem, seien es Gegenstände oder gar Lebewesen, ansprechen. Da heißt ein hölzerner Brotkasten „Krümelmonster", da trägt ein Bewegungsmelder den Namen „Mouse Watch". Oder ein ebenso praktischer wie eleganter Gürtelhänger reckt als „Gürtelschwan" majestätisch seinen gedoppelten Bügel-„Hals".

GAMES WITH SHAPES AND GAME WITH NAMES

Every design student should ideally have drunk in the fundamental rules of aesthetics with their mother's milk. An experienced designer such as Tassilo von Grolman not only knows these principles inside out, he loves to test them occasionally – without abandoning them altogether – and to refresh an existing shape with a free play of lines. His gift of making the most of all the senses, his own love of pleasure, his penchant for elegance and opulence cause him frequently enough to leave the sober construction of daily objects and to turn to things which are not absolutely necessary for life, but which enrich earthly existence tremendously and make it more refreshing. A parade of just a few of the designs such as the cigar ashtray, champagne cooler or récamiere betrays the personal lifestyle of the designer – someone who always heeds duty and simultaneously honours freedom; someone who always makes room for enjoyment for the eye, nose and taste buds alongside regulated daily work. It would, for example, be strange if the sporadic cigar smoker von Grolman thought up a utensil that was only useful for cigarette smokers and did not also have in mind the greatest comfort for himself and others like him for resting the cigar and knocking off the ash. A related contribution to design is the materialized admiration of the now legendary brass design by Bauhaus artist Marianne Brandt. Von Grolman paired Brandt's design from 1924 with a variation made of silver metal sheet. His "Hommage to Marianne Brandt" of which few copies were made, retains the principle of the original "ashtray with lid", namely the half covered ball section, and integrates it into a longer, extended shape. The contrast between the circular ash holder and the convenient handle for the smoker is revoked. In this way, von Grolman made a resting place for the portly relative of the cigarette in a comfortable bed. With his second cigar ashtray, the commercial product "Sancho", von Grolman takes pleasure in repeating the motif of rounding in the adjacent part and further extends it. The aluminium sand cast ashtray is always ready for service, and it is not by chance that it bears the name of the round, as-faithful-as-he-is-smart, squire-farmer Sancho Pansa, who had once assisted his master, the gaunt knight Don Quixote in Cervantes' grandiose novel, with humble devotion through all his adventures. Just as is the case with "Sancho", both sides of von Grolman – namely his practical sense of life as well as his fantasy-driven idealism – can be detected in the product descriptions of ulti-mately profane objects. The inventive, eloquent names contribute in so many ways to the vivid effect of the objects. They feed associations by appealing to the inherent hidden similarities to figurative objects or living things. Thus a wooden bread bin is called "Krümelmonster" (crumb monster), and a movement sensor alarm bears the name "Mouse Watch". Or a belt hanger, every bit as practical as it is elegant, called the "Gürtelschwan" (belt swan), stretches its "neck" or hanger majestically.

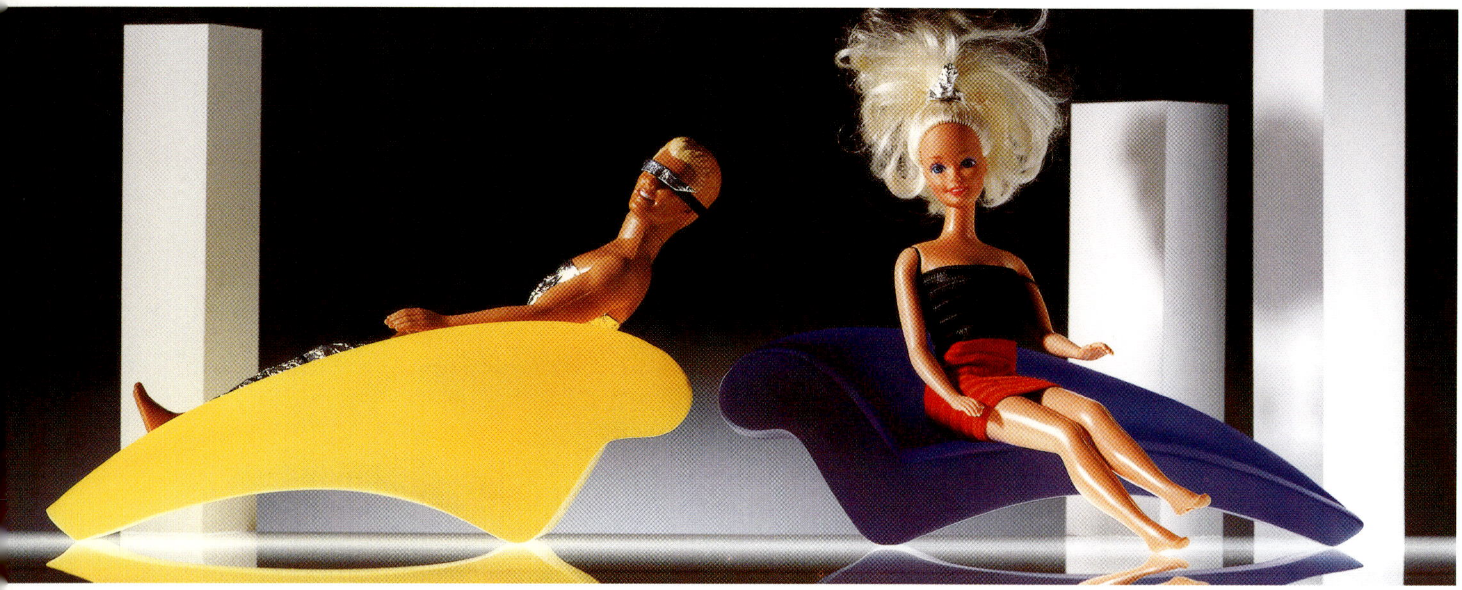

BILDSPRACHE UND SPRECHENDE BILDER

Apropos Eleganz: Ein Luxusgegenstand verlangt danach, in seiner Erscheinung und seinem Auftreten das Außergewöhnliche und Kostbare auch unwiderstehlich zur Schau zu tragen. Und in eben diesem Sinne hat Tassilo von Grolman der Kultfigur Barbie eine aufblasbare, popfarbene Récamiere zum Geschenk gemacht, deren klangvoll französischer Name allein schon eine Persönlichkeit verbürgt. Desgleichen hat er einen Champagnerkühler nicht nur entworfen, sondern ihn geradezu inszeniert. Auftraggeber Moët et Chandon machte seinem Büro zur Auflage, das Firmenzeichen sich kreuzender Schleifen mit dem Emblem der Firma in Erscheinung zu bringen. Von Grolman ließ also zunächst den Firmennamen in den konisch sich erweiternden Glaskörper einprägen. Der Schleife aber gab er plastischen Wert: Sie umfängt als eine Schärpe das gesamte Gefäß, überkreuzt sich beidseitig und dient an ihren höchsten Punkten dem Servierenden als willkommener Griff, womit ihr zugleich funktionale Bedeutung zukommt. Aus Metall gebildet wiederholt dieses dekorative wie praktische Schärpenelement gleichzeitig das Material von Fuß und Rand des Glaskörpers. Mit diesem Kunstgriff hat der Designer geschickt eine Auftragsarbeit zu seinem ureigenen Entwurf gesteigert. Der Clou aber ist die Innenbeleuchtung des Kühlers. Wenn nun die Eisstückchen wie Brillanten funkeln und der Champagnerkühler begierig wartet, dass man ihm die Flasche mit dem prickelnden Nass entnimmt, dann heißt es anzustoßen: Prosit dem Designer Tassilo von Grolman!

LANGUAGE OF IMAGES AND SPEAKING IMAGES

Apropos elegance: a luxury object also demands the open and irresistible display of the unusual and precious in its appearance and manner. And with just this in mind, Tassilo von Grolman gave the cult figure Barbie an inflatable, pop-coloured récamiere as a gift. The resonant French name alone already guarantees it a personality. In the same way, he not only designed a champagne cooler but also stage-managed it. Client Moët et Chandon instructed his office to make the logo with its crossing bows appear in the company emblem. So von Grolman first had the company name imprinted on the conical widening glass body. However, he gave the bow sculptural value: it surrounds the entire vessel as a sash, crosses over on both sides and serves at its highest point as a welcome handle for the pourer, also providing functional meaning. Formed from metal, this decorative as well as practical sash element repeats at the same time the material of the foot and rim of the glass body. With this trick, the designer has cleverly elevated a commissioned piece of work to his very own original design. The highlight is however the inner lighting of the cooler. When the ice cubes sparkle like diamonds and the champagne cooler eagerly awaits the removal of the bottle with its bubbling liquid, then it is time to raise our glasses: a toast to the designer Tassilo von Grolman!

Linke Seite
Aufblasbare Recamière als Geschenk für Barbie zum 35. Geburtstag. Im Rahmen des Projekts „Künstler und Designer gestalten für ein Kultobjekt", Mattel GmbH 1994.

Rechte Seite
Beleuchteter Champagner-Kühler „Starlight" für Moët et Chandon, 1999.

Left-hand side
Inflatable récamiere as a present for Barbie on her 35th birthday. Within the framework of the project "Artists and Designers design for a cult object", Mattel GmbH 1994.

Right-hand side
Lighted champagne cooler "Starlight" for Moët et Chandon, 1999.

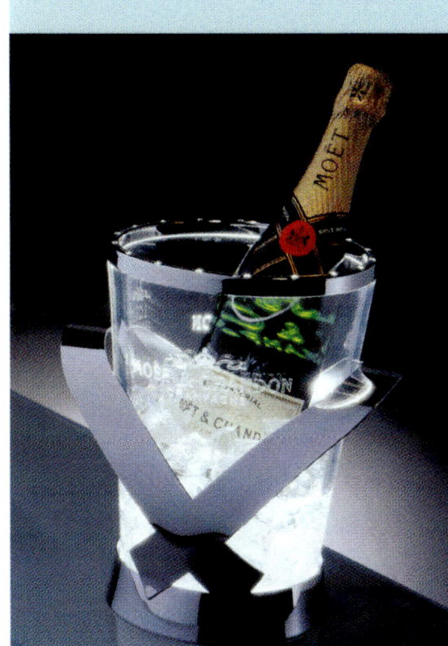

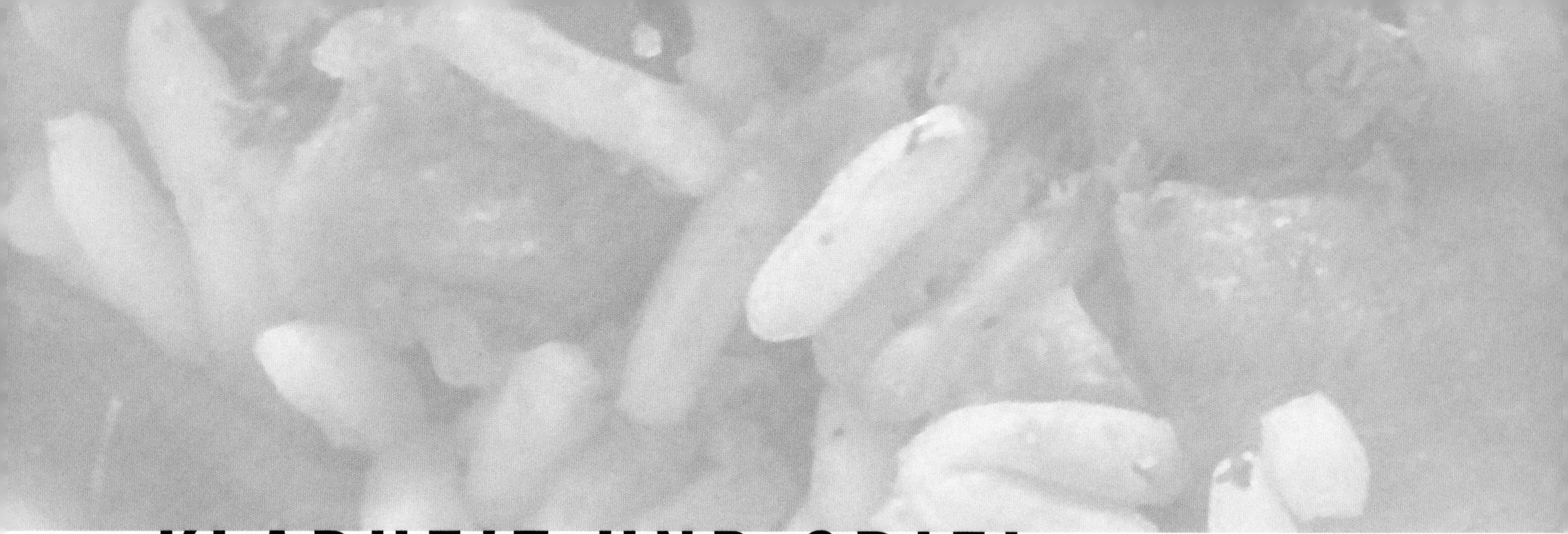

KLARHEIT UND SPIEL

2 Entenbrüstchen (je 200 g)
500 g Basmatireis
3 Knoblauchzehen
1 Bund Frühlingszwiebeln
frischer Koriander
4 EL Erdnussöl
150 g Sojasprossen
150 g Shitake-Pilze
1 Msp. Sambal
Salz
1 Msp. Chinagewürz
2 Würfel Hühnerbrühe
1 TL helle Sojasauce
0,2 l Weißwein

ENTENBRUST MIT SOJASPROSSEN UND SHITAKE-PILZEN

1 Die Brüstchen salzen, mit dem Chinagewürz einreiben und mit einem Zahnstocher kleine Löcher in die Haut – aber nicht ins Fleisch – stechen. Dann bei 200 °C ungefähr 20–30 Minuten im Backofen rosa braten.

2 Den Reis waschen, in den Topf geben, fingerdick mit Wasser bedecken und zugedeckt ungefähr 20 Minuten auf kleinster Flamme köcheln lassen.

3 Den Knoblauch schälen und klein hacken. Die Frühlingszwiebeln schälen und in Ringe schneiden. Den Koriander klein hacken. Die Shitake-Pilze waschen und in feine Scheiben schneiden.

4 Die rosa gebratenen Entenbrüstchen aus dem Ofen nehmen und in ungefähr 1 cm dicke Scheiben schneiden.

5 Das Erdnussöl in einem Wok erhitzen, den Knoblauch, die Frühlingszwiebeln und die Sojasprossen zugeben und kurz anbraten. Ungefähr 1–2 Minuten köcheln lassen und alles ein paar Mal wenden. Die gebratenen Entenbrüstchen, die Shitake-Pilze, die Hühnerbrühe, die Sojasauce, den Sambal und den Wein zufügen, den Koriander darüber streuen und mit dem Reis servieren. Dazu passt ideal ein kalter Rosé.

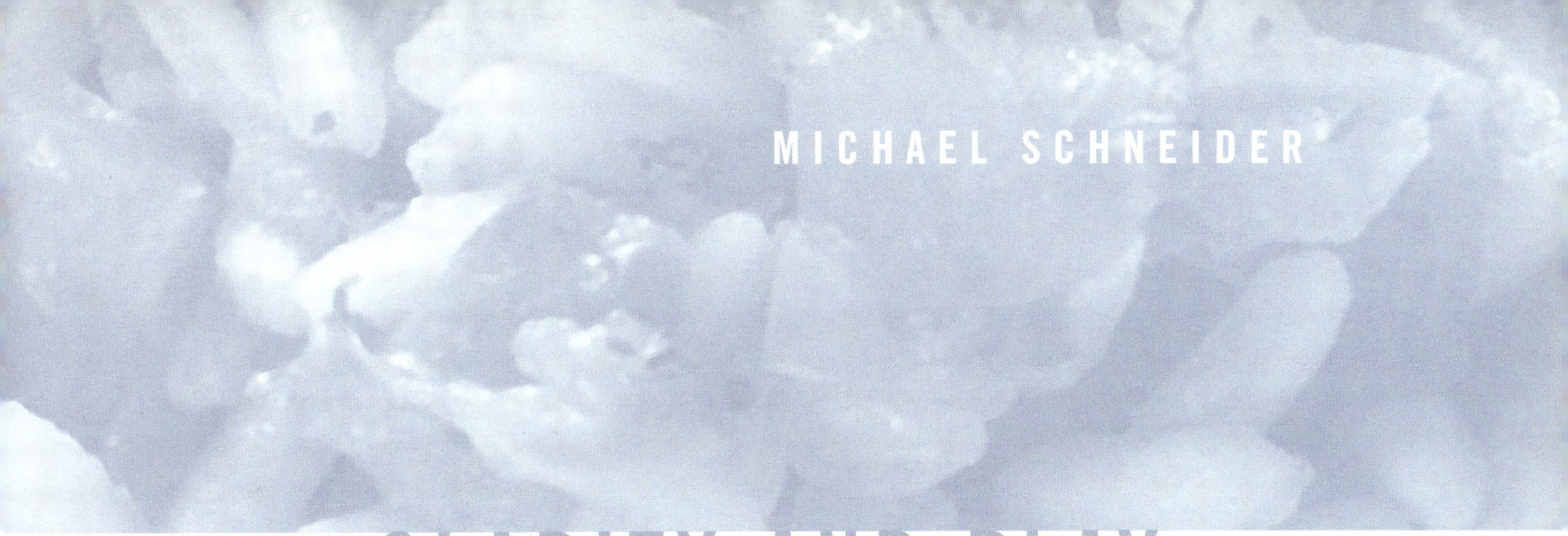

MICHAEL SCHNEIDER

CLARITY AND PLAY

BREAST OF DUCK WITH BEAN SPROUTS AND SHITAKE MUSHROOMS

1 *Salt the duck breast, rub with the Chinese spices and with a toothpick make small holes into the skin but not in the flesh. Then roast in the oven at 200 °C for around 20–30 minutes until pink.*

2 *Wash the rice, put into a pan, cover with a finger depth of water and allow to simmer for around 20 minutes on the lowest heat.*

3 *Peel the garlic and chop finely. Peel the spring onions and cut into rings. Chop the coriander finely. Wash the shitake mushrooms and slice thinly.*

4 *Take the roasted duck breasts out of the oven and cut into slices approximately 1 cm thick.*

5 *Heat the peanut oil in a wok, add the garlic, spring onions and the bean sprouts and fry briefly. Allow to simmer for 1–2 minutes, turning a few times. Add the roasted duck breasts, the shitake mushrooms, the sambal, the chicken stock, the soy sauce and the wine. Sprinkle the coriander over it and serve with the rice. A cold rosé is ideal with this dish.*

2 duck breasts (each 200 g)

500 g basmati rice

3 cloves of garlic

1 bunch spring onions

Fresh coriander

4 tbsp peanut oil

150 g bean sprouts

150 g shitake mushrooms

1 pinch sambal

Salt

1 pinch Chinese spices

2 cubes chicken stock

1 tsp light soy sauce

0,2 l white wine

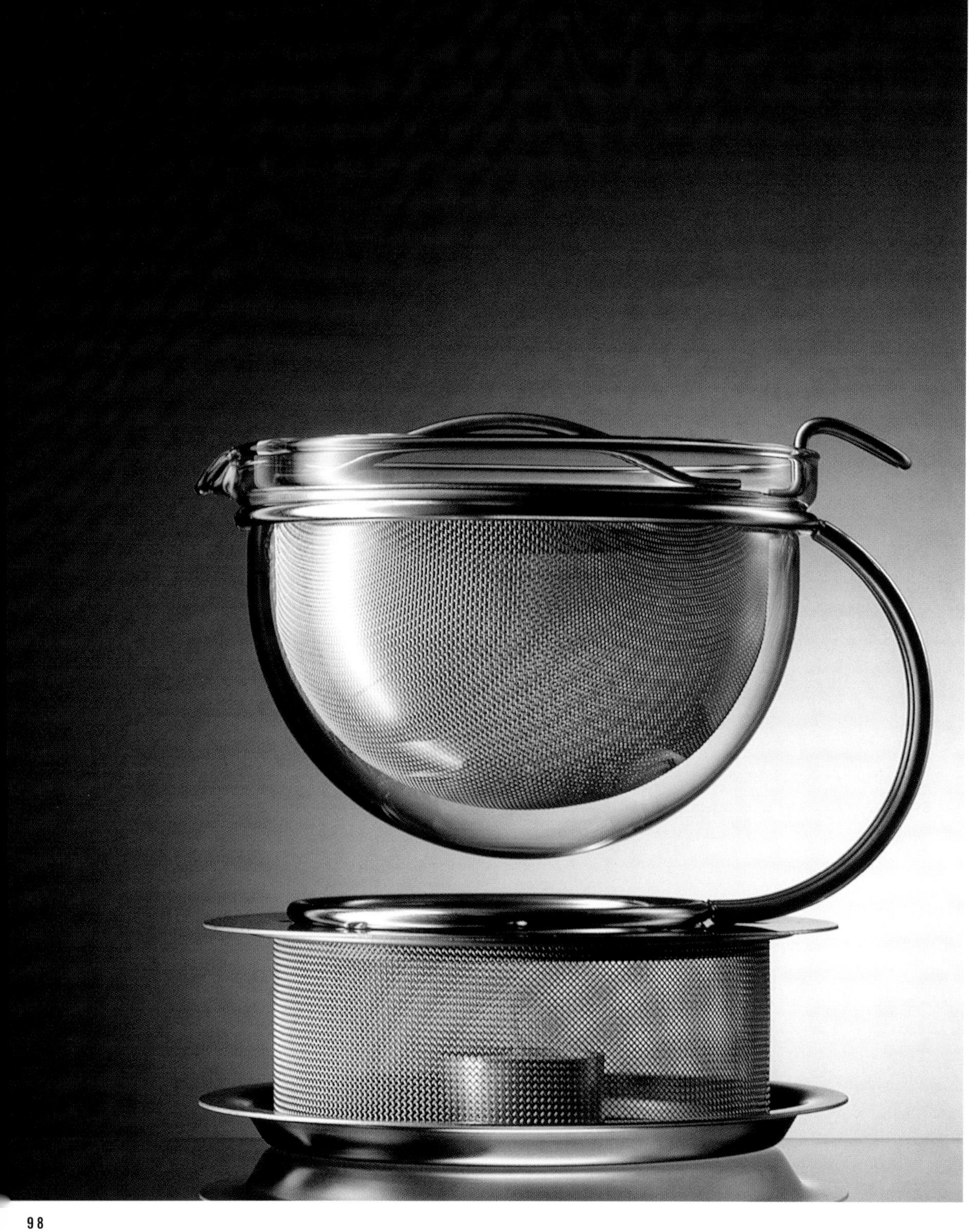

Teekanne „Mono Filio“ mit Rechaud (1993) für Mono Metallwarenfabrik GmbH, 1998.

Teapot “Mono Filio” with réchaud (1993) for Mono Metallwarenfabrik GmbH, 1998.

ERSTER BLICK: DER ZEREMONIENMEISTER

Tee wird traditionell im Dunkeln geboren: Im Inneren undurchsichtiger Gefäße aus Porzellan, Keramik oder Metall. Die Teeblätter werden in Trichtern oder – schlimmer noch – in kleine Teeeier gedrückt ins Wasser gesenkt und nach genau bemessener Zeit wieder aus dem Bauch der Kanne herausgeholt. Die Geburt des Tees war ein Vorgang im Unsichtbaren. Tassilo von Grolmans Kannen zeigen, wie Tee wird. Dort schwimmen Teeblätter frei, fast im ganzen Volumen der Kanne. Alles ist transparent. Das Spiel der Strömungen, der Wirbel, das langsame, pendelnde Absinken der Teeblätter, die Färbungen des Wassers – Faszination für die Augen. Von Grolman erweiterte die Zeremonie des Teetrinkens um die Zeremonie des Teewerdens. Woher kennen wir diese Faszination, wenn wir die Veränderung bei der Begegnung und Verschmelzung von unterschiedlichen Elementen beobachten? Richtig, aus dem Labor. Glaskolben und Bunsenbrenner. Tassilo von Grolman hat also ein Tee-Labor gebaut, formales wie materielles Repertoire stammen von dort. Später folgte auch ein Kaffee-Labor nach, bei dem das Kerzenlicht die Färbung und die changierende Farbe des Kaffees zum visuellen Genuss macht. Soweit: typisch Techniker – Tassilo von Grolman hat Maschinenbau studiert. Es gibt da aber noch etwas ganz Anderes zu sehen, nämlich ein poetisches Gebilde aus verschiedenen Materialien, Strukturen, Verformungen, das bei strenger und scheinbar rein funktionaler Bearbeitung Freiräume des Spielerischen mit einfängt. Das erzählt vom Designer Tassilo von Grolman, dem Zeremonienmeister, der Weste trägt. Welche Weste man trägt, ist eine Design-Frage. Wie man sie trägt, ist ein poetisches Geheimnis. Nachahmer könnten noch nicht einmal eine Zitrone gewinnen.

FIRST GLANCE: THE MASTER OF CEREMONIES

Tea is traditionally born in the dark: in the inside of an opaque vessel made of porcelain, pottery or metal. The tea leaves are pressed into funnels or – even worse – into little tea infusers, dropped into water and retrieved from the bowels of the teapot after a precisely measured time. The birth of the tea was an invisible event. Tassilo von Grolman's teapots show how tea becomes tea. There the tea leaves swim freely, in almost the entire volume of the teapot. Everything is transparent. The play of the currents, the swirling, the slow oscillating sinking of the tea leaves, the colouring of the water – fascination for the eyes. Von Grolman added the ceremony of tea creation to the ceremony of tea drinking. Why is this fascination with observing the change that occurs at the meeting and fusion of different elements familiar? Correct – we know it from the laboratory. Glass retorts and Bunsen burners. So Tassilo von Grolman has built a tea laboratory, from which formal and physical repertoires originate. Later a coffee laboratory followed, in which candlelight transforms the hue and the changing colour of the coffee into visual pleasure. So far: typical of a technician – Tassilo von Grolman had studied engineering. There is, however, still something completely different to be seen here, namely a poetic sculpture made of various materials, structures, forms, which also captures the freedom of the playful in its strict and seemingly purely functional handling. This tells us about the designer Tassilo von Grolman, the master of ceremonies, who wears waistcoats. Which waistcoats are worn is a question of design. How they are worn, is a poetic secret. Imitators could not hold a candle to him.

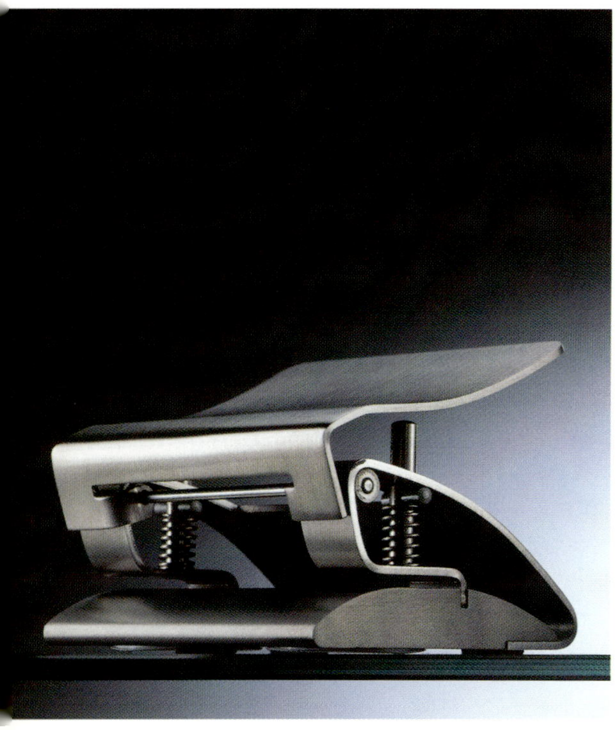

ZWEITER BLICK: DER MATERIALEXPERTE

Als der Computer die Chefbüros eroberte, war plötzlich das Meisterstück von Montblanc wieder da. Der Füller, ein fast schon vergessenes Relikt aus einer Zeit, der ein Name wie „Tassilo“ entstammt. Dem digitalen Austausch von Buchstaben stellt Montblanc die persönliche Handschrift entgegen. Worte sind hier also nicht als Aneinanderreihung von Buchstaben, sondern als visuelle Gesten gedacht, wie sie eine Hand mit einer bestimmten Feder auf ganz bestimmtes Papier formt. Eine Schreibtisch-Garnitur zu gestalten, die vollständig dem Schreiben und dem Papier gewidmet ist, verführt auf der Schwelle zum 21. Jahrhundert leicht zur Nostalgie. Was aber macht Tassilo von Grolman? Er faltet Metall. Lauter Metallblätter. Eine Hommage aus Metall an das Papier. Der Name ist Programm: „Steel“. Und wie immer findet der Herr mit der Weste klare, einfache, „logische“ Formen. Wie immer: sichtbare, einsichtige Funktionen. Wie immer auch: das Spiel mit Licht und Schatten der Oberflächen und der Verformungen. Wer Nostalgie erwartet hat, sieht Modernität. Wer noch mehr Nippes für den Schreibtisch ahnte, sieht Formen von strenger Disziplin. Wer sich formalen Gewichtigkeitspomp zum schwellenden Chefsessel vorstellte, sieht spielerische Leichtigkeit. Wie kommt bei soviel Klarheit das Spielerische hinein? Seine Utensilien für die Kulturzone Schreibtisch haben weder Dekor noch Farbigkeit. Es muss der Gourmet sein, der aus einfachen Zutaten einen so reichen Genuss zubereiten kann. Tassilo von Grolman kocht gern für seine Gäste. Die Krawatte wird beim Kochen genau zwischen dem dritten und vierten Knopf in das Hemd gesteckt.

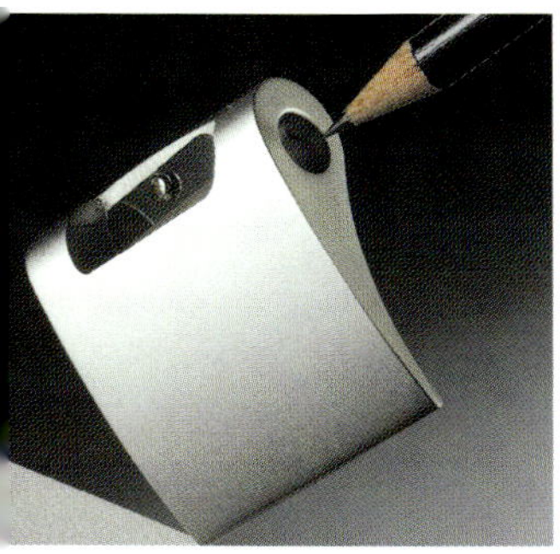

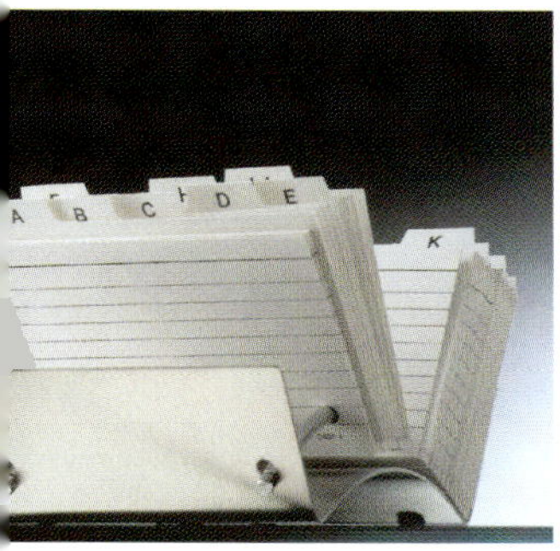

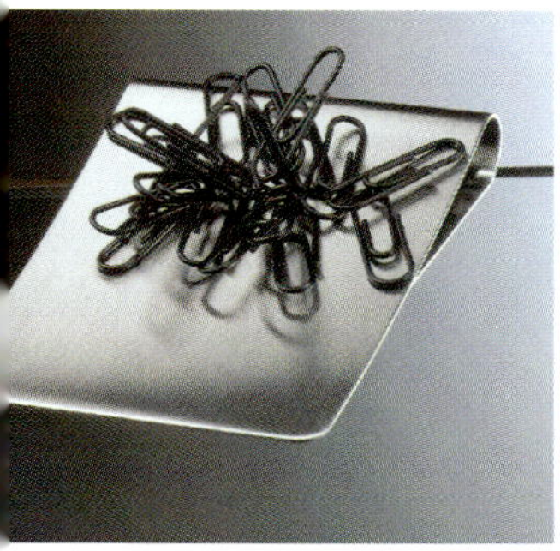

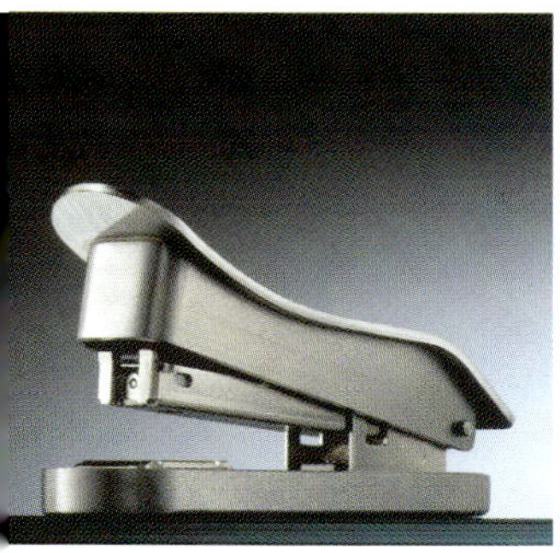

SECOND GLANCE: THE MATERIALS EXPERT

When the computer conquered the bosses' offices, Montblanc's "Masterpiece" suddenly reappeared. The fountain pen, an almost forgotten relic from a time that generated a name like "Tassilo". Montblanc sets personal handwriting against the digital exchange of letters. So here words are not intended as an enumeration of letters but as visual gestures, as formed by a hand with a particular quill on a particular paper. To design a desk set which is entirely devoted to writing and paper encourages nostalgia on the threshold of the 21st century. But what does Tassilo von Grolman do? He folds metal. Nothing but sheets of metal. A homage to paper made of metal. The name of the range is: "Steel". And as always the man with the waistcoat finds clear, simple, "logical" designs. As always: visible, understandable functions. And as always: the play of light and shade of the surfaces and the distortions. He who expected nostalgia sees modernity. He who foresaw still more knick-knacks for the desk sees designs of strong discipline. He who imagined formal heavy pomp for the swelling boss's chair sees playful lightness. Where does playfulness come into so much clarity? His utensils for the culture zone of the desk have neither decorations nor colourfulness. It must be the gourmet who can prepare such rich pleasure from simple ingredients. Tassilo von Grolman enjoys cooking for his guests. When cooking, his tie is tucked into his shirt exactly between the third and fourth button.

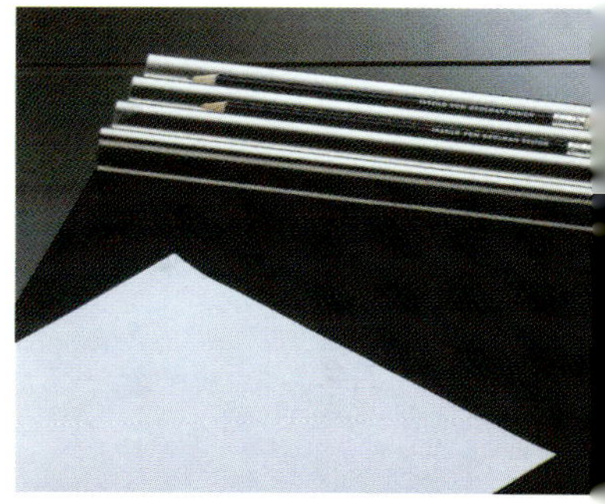

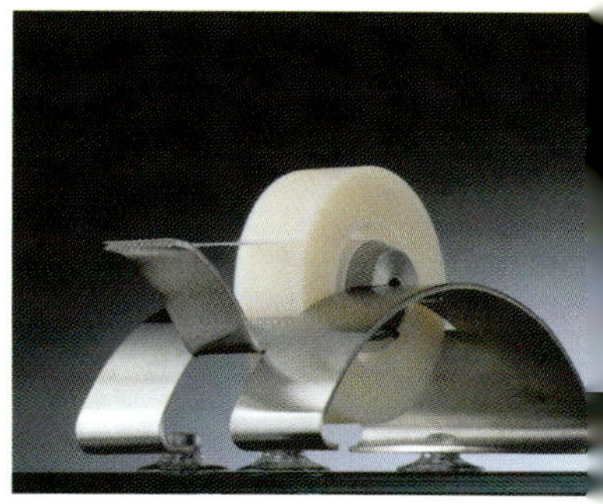

Büroaccessoires-Serie „Steel" für Emform GmbH, 1999.

Office accessory series "Steel" for Emform GmbH, 1999.

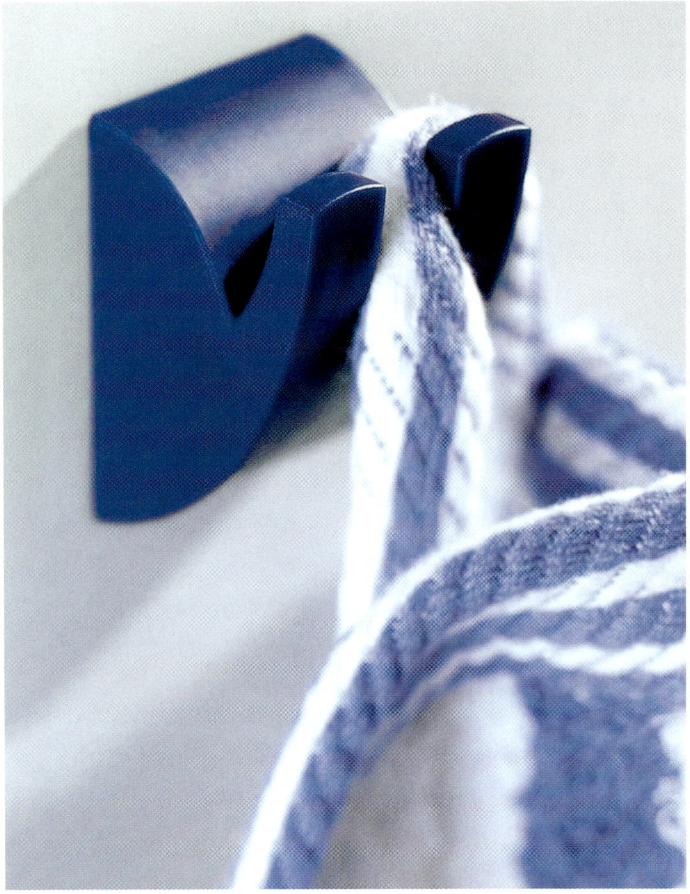
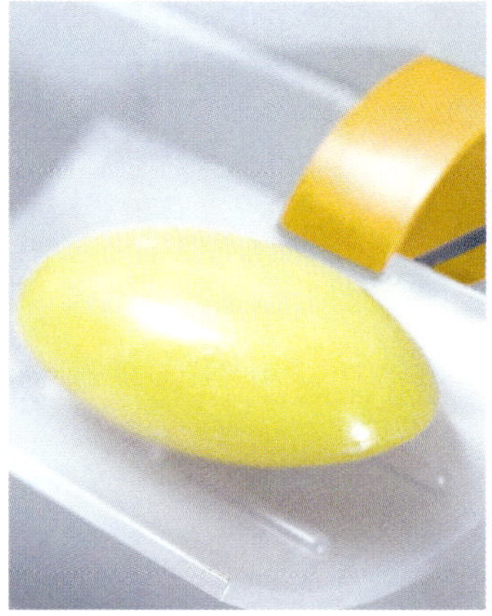
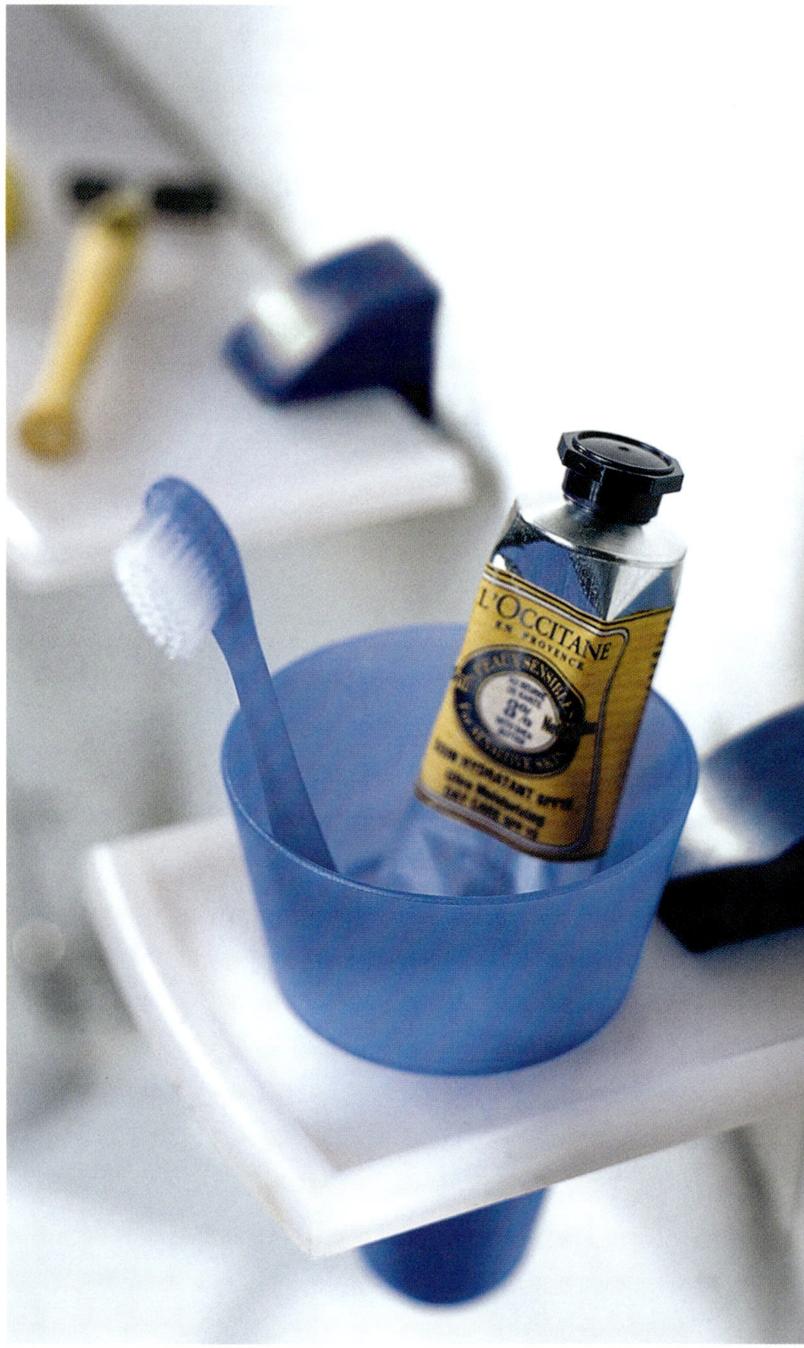
L'OCCITANE
EN PROVENCE
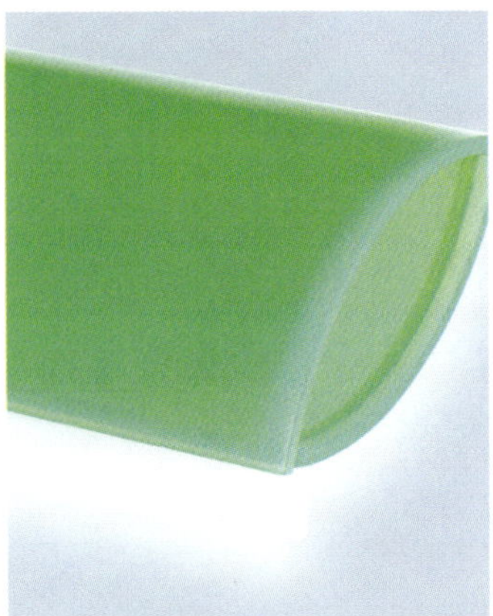
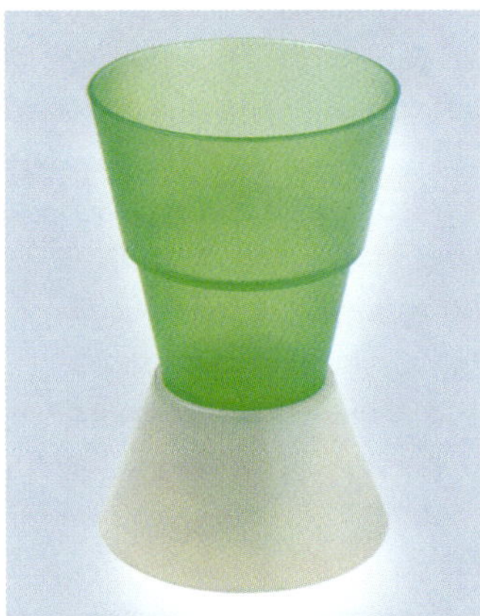
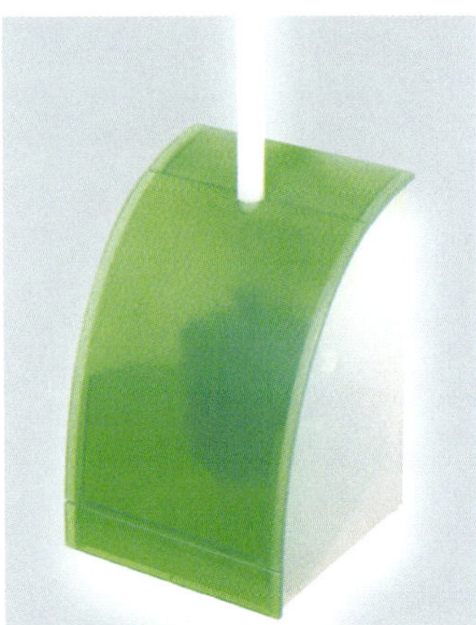
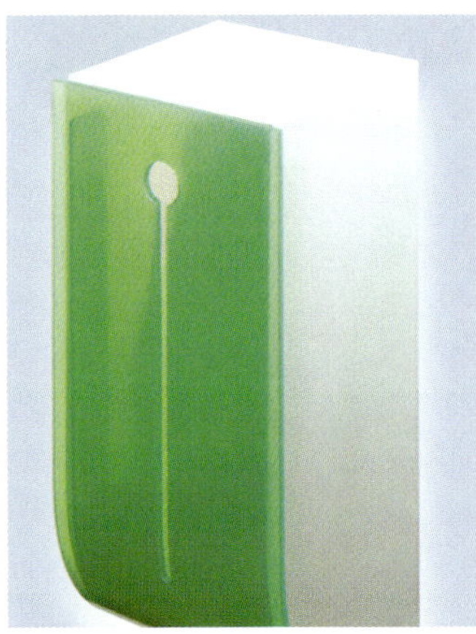

DRITTER BLICK: FRA DESIGNICO

Eine der schwierigsten Aufgaben in der Hohen Schule der Schauspielkunst, Cary Grant hätte einen Oscar dafür verdient, ist, wie man zerknittert, zerzaust und zerknautscht morgens ins Bad geht und sich das erste Mal im Spiegel sieht – und dabei die Frauenherzen schwach macht. Die meisten Frauen möchten die erste halbe Stunde im Bad allein sein. Dann erst hat der Rest der Familie Zutritt. Tassilo von Grolman, den sich auch gute Freunde weder zerknittert, noch zerzaust oder zerknautscht vorstellen können, schafft optische Zutaten für die Ausstattung unserer Badezimmer. Er will uns bei unserer ersten täglichen Begegnung mit uns selbst Beistand leisten. Er malt für uns luftige, zarte Farbtupfer in Aquarell, wattig, wolkig, wie Töne. Nichts von Waschtrog-Atmosphäre oder steriler Hygiene, „Lavela" ist eine offene Aufforderung zum Spiel mit Wasser und Düften, zur guten Laune, zu zweit oder zu dritt. Auch hier gibt es kein Dekor, höchstens die Löcher in den Seifenschalen, aber auch sie sind streng geometrisch. Statt Ornamente gibt es Gläser als Kelche – und Farbe, Transparenz und Licht. Wer noch nicht gleich in den Spiegel schauen mag, findet woanders freundliche Gegenüber. Dass dahinter ein durchdachtes technisches System von Haltern aus wenigen Teilen steckt, merkt der Kenner wohl. Uns, den Genießern, gefällt, dass es ein System ist, das keine systematischen Zwänge der Anordnungen erzeugt, sondern uns die Freiheit des Nebeneinander, Übereinander und – wenn wir wollen – auch Durcheinander lässt. Man kann sich vorstellen, dass in einem solchen Bad sogar am frühen Morgen schon mal vorsichtig gelacht wird. Und dass der erste, der ins Bad geht, die Tür hinter sich offen lässt.

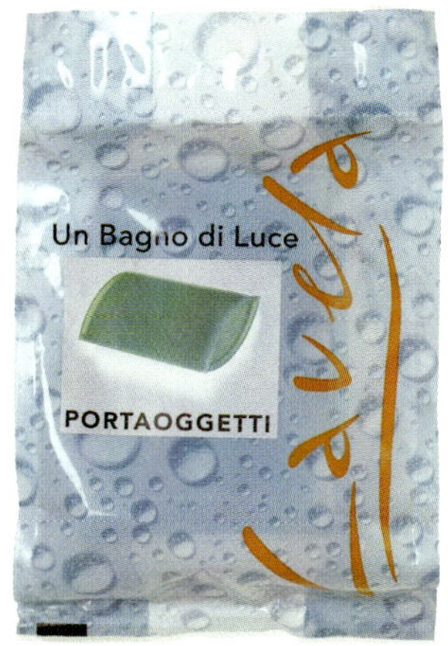

THIRD GLANCE: FRA DESIGNICO

One of the most difficult exercises at the High School of drama – Cary Grant would have deserved an Oscar for it – is, how to go to the bathroom in the morning, crumpled, dishevelled and worse for wear and take your first look in the mirror, and still make women's hearts flutter. Most women want to spend the first half hour in the bathroom alone. Only then is the rest of the family allowed in. Tassilo von Grolman, whom even good friends can imagine neither crumpled, dishevelled nor worse for wear, creates optical ingredients for the fitting out of our bathrooms. He wants to lend support to us for our first daily meeting with ourselves. He paints for us airy, soft coloured dots in watercolour, cotton woolly, cloudy, like notes. None of that washroom atmosphere or sterile hygiene, "Lavela" is an open invitation to play with water and scents, to a good mood, as a couple or as three. There is no decoration here either, at most the holes in the soap dishes, but they are also strictly geometric. Instead of ornaments there are glasses as goblets – and colour, transparency and light. He who does not yet want to look immediately in the mirror will find something friendly confronting him elsewhere. The connoisseur will probably notice that a well thought-out technical system of holders made of few parts is behind it. What we – those who just enjoy it – like about this system is that it does not systematically enforce order but allows us the freedom to leave things next to each other, on top of each other and also – if we want – in a muddle. You can imagine that there is already cautious laughter in such a bathroom even early in the morning. And also that the first into the bathroom leaves the door open behind them.

Linke Seite
„Lavela", transluzenter Kunststoff in den Farben Grün, Blau, Rot, Gelb und Aquagrau in Kombination mit mattsilberfarbigen Metallteilen, seit 1999 auf dem Markt. Die Serie wurde für SB- und Baumärkte konzipiert.

Rechte Seite
Tassilo von Grolman entwarf die Verpackungen für „Lavela" im Team mit Peter Engelhardt für Plaudo Italia.

Left-hand side
"Lavela", translucent plastic in the colors green, blue, red, yellow and aqua gray in combination with matte silver metal parts, on the market since 1999. The series was designed for DIY and hardware stores.

Right-hand side
Tassilo von Grolman designed the packaging for "Lavela" in cooperation with Peter Engelhardt for Plaudo Italia.

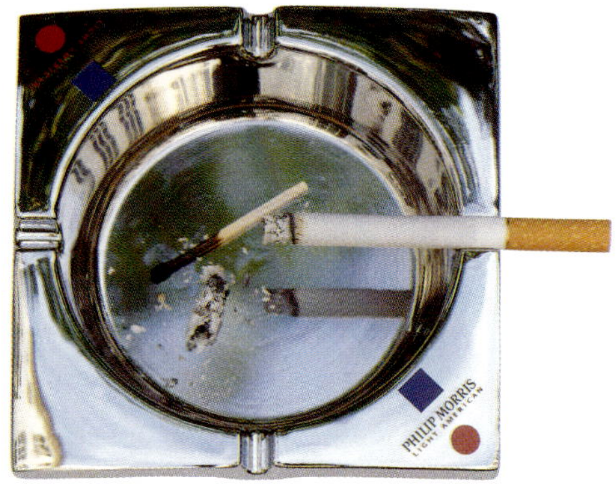

Linke Seite
Give-aways für die Gastronomie.
Philip Morris Light American, 1996.

Rechte Seite
Zigarrenaschenbecher „Sancho" für Troika, 1998.

Left-hand side
Give-aways for gastronomy, Philip Morris Light American, 1996.

Right-hand side
Cigar ashtray "Sancho" for Troika, 1998.

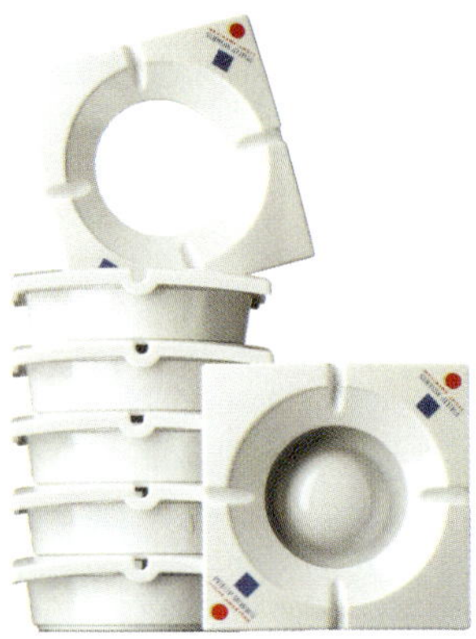

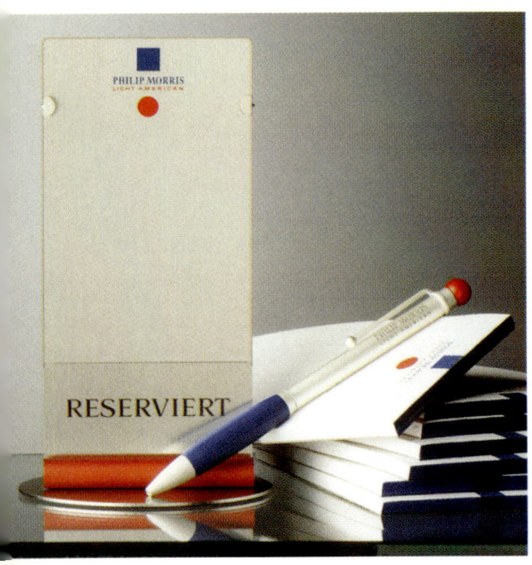

VIERTER BLICK: DER UNTERHÄNDLER

Wie ist das eigentlich in den USA: Kann man da noch jemanden unbefangen und unbestraft auffordern, die Friedenspfeife zu rauchen? Und wie ist das unter Gentlemen: Wird ein Westenträger dem anderen den Genuss einer guten Zigarre untersagen oder ihn auf den Balkon schicken? Tassilo von Grolman hat sein Geständnis visuell abgelegt. Für die Zigarre schuf er eine richtiggehende Huldigung in zwei Varianten. „Hommage à Marianne Brandt" und „Sancho" sind rituelle Gefäße für die Refugien, die den Genießern des blauen Dunstes noch bleiben. Beide opulent im Material, beide kosten das Spiel von Spiegelungen, Reflexen und Schattierungen voll aus. Ein Bekenntnis des Genusses, der Lebensfreude, der wohlbedachten Luxuriosität. Und das erste Mal ein deutliches Blinzeln – eine organische Form und eine erotische Anspielung. Für die Zigarettenraucher in größerer Gesellschaft stapelt von Grolman einen Turm. Jedes einzelne Teil bietet vier Chancen, den Teppich des Gastgebers zu schonen. Eigentlich sollte seine Form die Unterlage für Künstler-Dekore einer Philip-Morris-Edition sein. In seiner puren Form ist sie jedoch das wichtigste Sammlerstück.

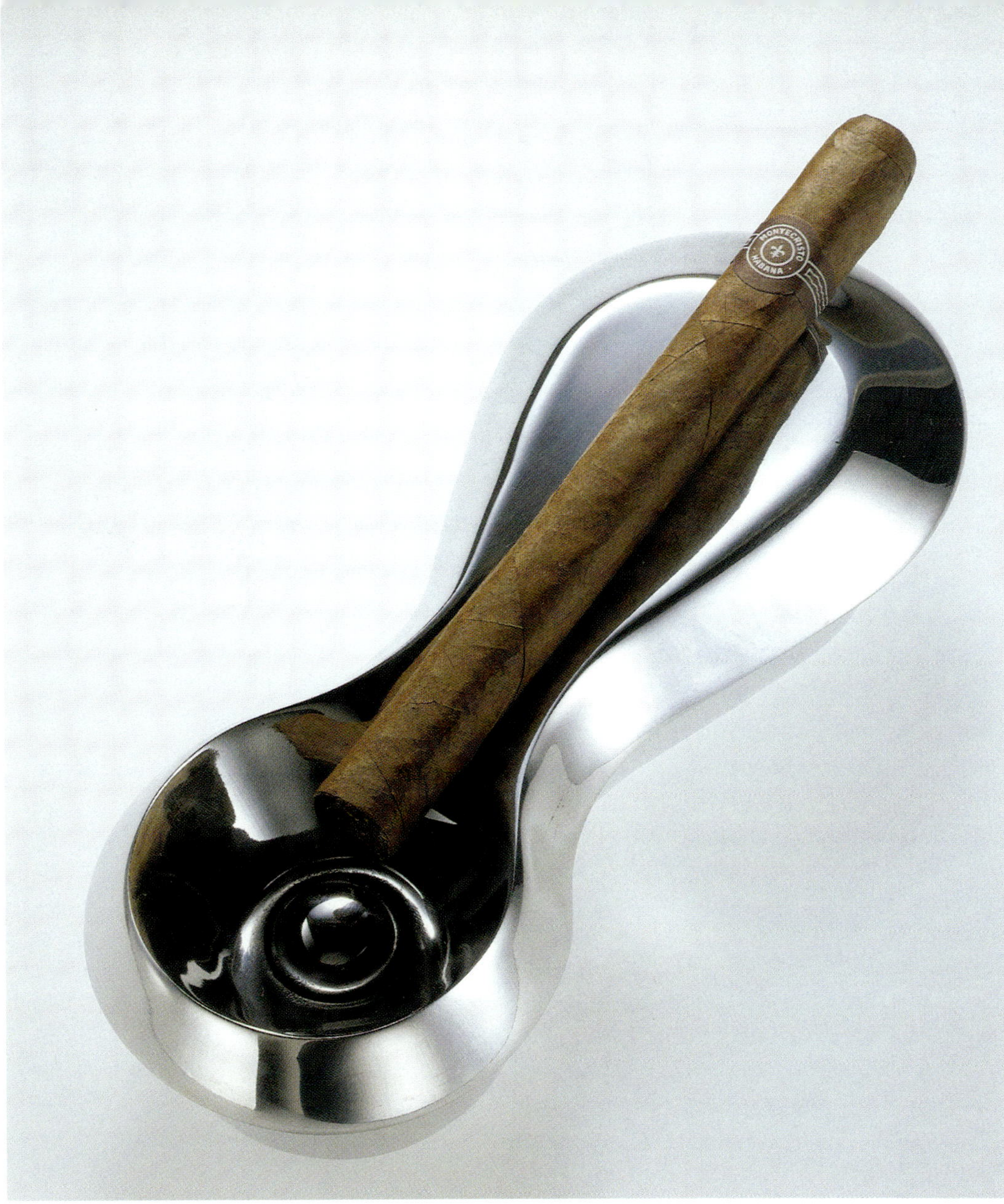

FOURTH GLANCE: THE NEGOTIATOR

What is it actually like in the USA: is it still possible to ask somebody uninhibitedly and unpunished to smoke the pipe of peace? And how does it work amongst gentlemen: will one waistcoat-wearer prevent the other from enjoying a good cigar or send him onto the balcony? Tassilo von Grolman has made a visual confession. For the cigar, he created a genuine homage in two variations. "Hommage à Marianne Brandt" and "Sancho" are ritual vessels for the remains that are all that is left to those who enjoy the blue haze. Both opulent in their materials, both make the most of the play of reflections, reflexes and shades.
A confession of pleasure, of joy of life and well-considered luxury. And for the first time an obvious wink – an organic shape and an erotic allusion. For the cigarette smoker in wider society, von Grolman erects a tower. Each individual part offers four opportunities to spare the host's carpet. The shape should really be the basis for artists' decoration of a Philip Morris edition. In its pure form it is however the most important collector's piece.

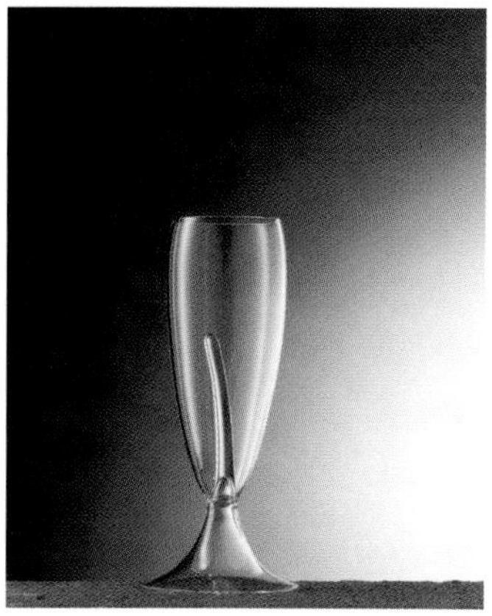 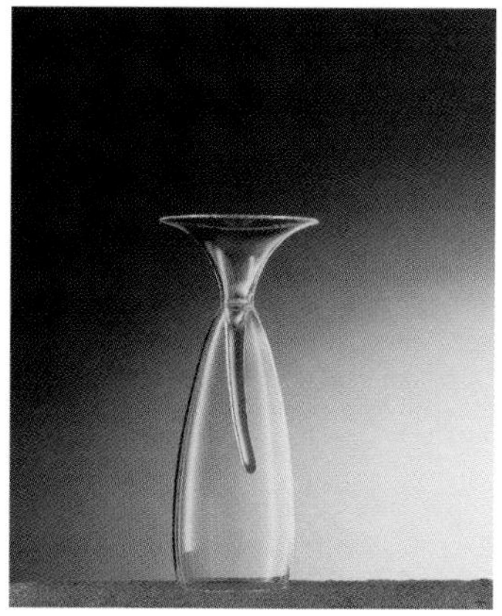 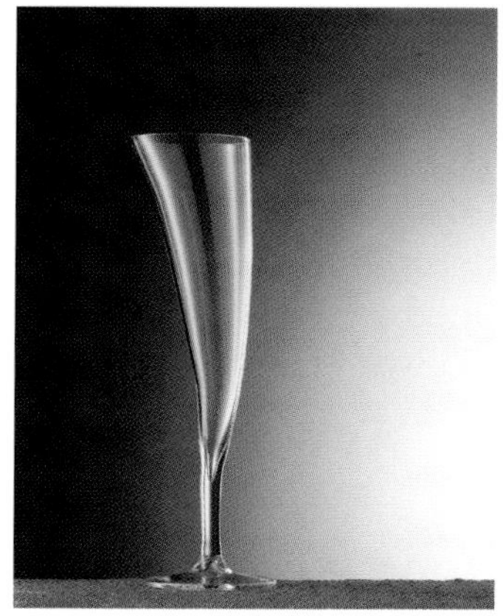

FÜNFTER BLICK: DER DIRIGENT

Bei diesem fünften Blick muss man dreimal hinsehen.

Einmal: Tassilo von Grolman dekoriert Gläser. Der Herr mit der Weste als Dekorateur? Die Partybecher der fünfziger Jahre mit ihren anonymen Kreis-, Karo-, Blümchen-, Spielkarten-, Streifen- und Spiralmustern werden von Ritzenhoff als Marketingidee wiederbelebt – auf höherem Niveau versteht sich. So sind Massenprodukte denn durch austauschbare Dekorteile „individualisiert", Grundformen werden nämlich durch „Künstler-Editionen" dekoriert – Marketing-Ideen, keine ursprünglichen Design-Ideen. Ich bin da zwiefach befangen. Zum einen empfinde ich die anonymen Partybecher der fünfziger Jahre als authentischer als die heutigen Künstler-Dekor-Editionen. Zum anderen finde ich Tassilos Dekore schön – außerdem hat er mir immer eins geschenkt. Nein, Ornament ist kein Verbrechen. Kunst am Bau ist ja manchmal auch schon gut gegangen.
Zweimal: Tassilo von Grolman, der Techniker, Handwerker und poetische Gourmet, hat bei Lauscha das erste Mal nicht Serien-Ingenieure, sondern Handwerker als Partner – die Glasbläser. Jenes Zusammenspiel von Idee und handwerklichem Können ist hier der Ausgangspunkt für Design. Und spontan wird hier etwas frei, was von Grolman sonst erst nach systematischer Durchdringung zu seinen klaren Formen und seinem spielerischen (poetischen) Gehalt bringt: Der vom systematischen Denken und Konstruieren geprägte Maschinenbauer zeigt, wie emotional er aufgeladen ist, wie sensibel und akrobatisch er die freien Kräfte von Materialeigenschaften und die Fähigkeiten der Glasbläser zu Formen dirigieren kann, deren poetische Qualität nicht „konstruierbar" und von Maschinen nicht reproduzierbar ist.
Dreimal: Zufällig treffe ich auf einer der Kreuzungen der Messegänge mit Tassilo von Grolman zusammen. Er stellt seine schweren Taschen ab, begrüßt mich und legt los. Unter der Weste die geschwellte Brust voll von bunten Sternchen, vollgepumpt mit allem, was er gesehen hat und was ich unbedingt auch sehen muss. Ein großer Junge mit weißem Haar, der seine Weste trägt wie keiner, hoch motiviert, ansteckend gut gelaunt und voller Begeisterung. Einer der sehr wenigen Designer, die sich für die Arbeit anderer Designer begeistern können. Das Klare und das Spielerische, wie ist das bei ihm zu erklären? Das Klare klärt bei ihm etwas, das nämlich, wozu die Dinge eigentlich da sind. Das Spielerische aber nährt: Seine Dinge sind für Genießer gemacht. Und die sind – das weiß er nur zu gut – wie Kinder.

Linke Seite
1999 war Weimar Kulturstadt Europas. Aus diesem Anlass entwarf von Grolman für die Designer-Collection des Herstellers Lauscha-Glaskunst die Champagnergläser „Iphigenie" und „Christiane". Durch einfaches Wenden wird „Christiane" zu einem Schnapsglas.

Rechte Seite
Dekore für Ritzenhoff Collections. Wasserglas (1999), Milchglas (1992), Teeglas und Teedose (1999).

Left-hand side
In 1999, Weimar was "culture city" of Europe. On this occasion, von Grolman created the champagne glasses "Iphigenie" and "Christiane" for the Designer-Collection of the manufacturer Lauscha Glaskunst. "Christiane" turns into a schnapps glass simply by turning it over.

Right-hand side
Decors for Ritzenhoff Collections. Water glass (1999), milk glass (1992), tea glass and tea caddy (1999).

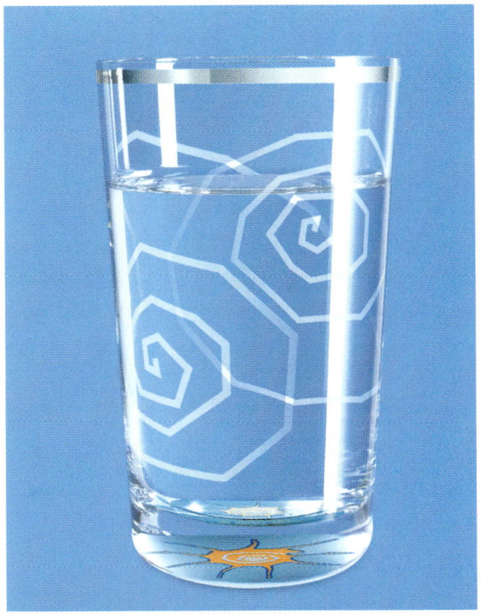

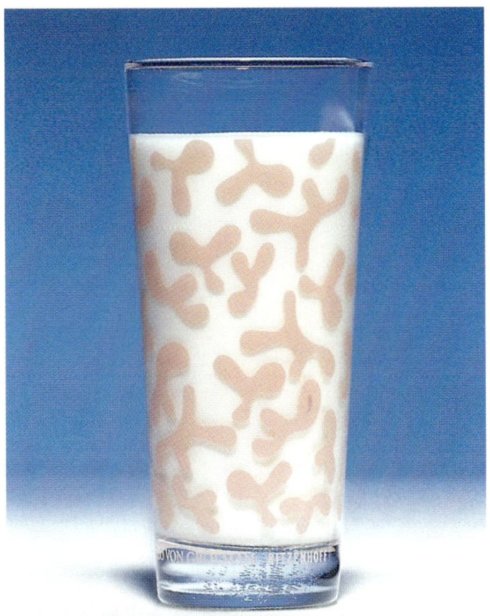

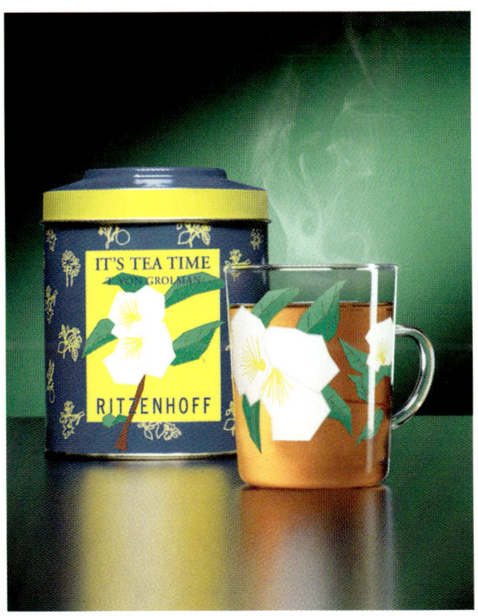

FIFTH GLANCE: THE CONDUCTOR

At this fifth glance, you have to look three times.

Firstly: Tassilo von Grolman decorates glasses. The man with the waistcoat as a creator of decoration? The party cups of the fifties with their anonymous circular, diamond, flower, playing card, striped and spiral patterns are resuscitated by Ritzenhoff as a marketing idea – on a higher plane, of course. So mass-produced objects are "individualized" through exchangeable decorative parts, basic shapes are thus decorated using "artist editions" – marketing ideas, not original design ideas. I am doubly prejudiced here. On the one hand, I feel that the anonymous party cups of the fifties are more authentic than the artist's decoration edition of today. On the other hand, I think that Tassilo's decorations are beautiful – furthermore, he has always given me one. No, ornament is no crime. Art in construction has already even turned out well on occasion.
Secondly: Tassilo von Grolman, the technician, craftsman and poetic gourmet, has for the first time not serial engineers but craftsmen as partners at Lauscha – the glass blowers. The starting point for design here is the interplay of idea and craftsmanship. And spontaneously something is set free here, something that von Grolman otherwise brings to his clear shapes and his playful (poetic) content only after systematic exploration. The engineer – moulded by systematic thinking and construction – shows how emotionally charged he is. He shows how sensitively and acrobatically he can direct the free energies of the materials' properties and the skills of the glass blowers to create shapes, the poetic quality of which is not "constructible" and cannot be reproduced by machines.
Thirdly: I meet Tassilo von Grolman by chance at one of the crossroads of the exhibition halls. He puts his heavy bags down, greets me and starts talking. Under the waistcoat, a swollen chest full of colourful stars, fired up with everything he has seen and what I must definitely also see. A big lad with white hair, who wears his waistcoat like no other, highly motivated, in infectiously good humour and full of enthusiasm. One of the very few designers who can be enthusiastic about the work of other designers. The clear and the playful, how can that be explained with him? Clarity clarifies something for him, namely what the things are actually there for. But the playful nourishes: His things are made for pleasure lovers. And they are – he knows this only too well – like children.

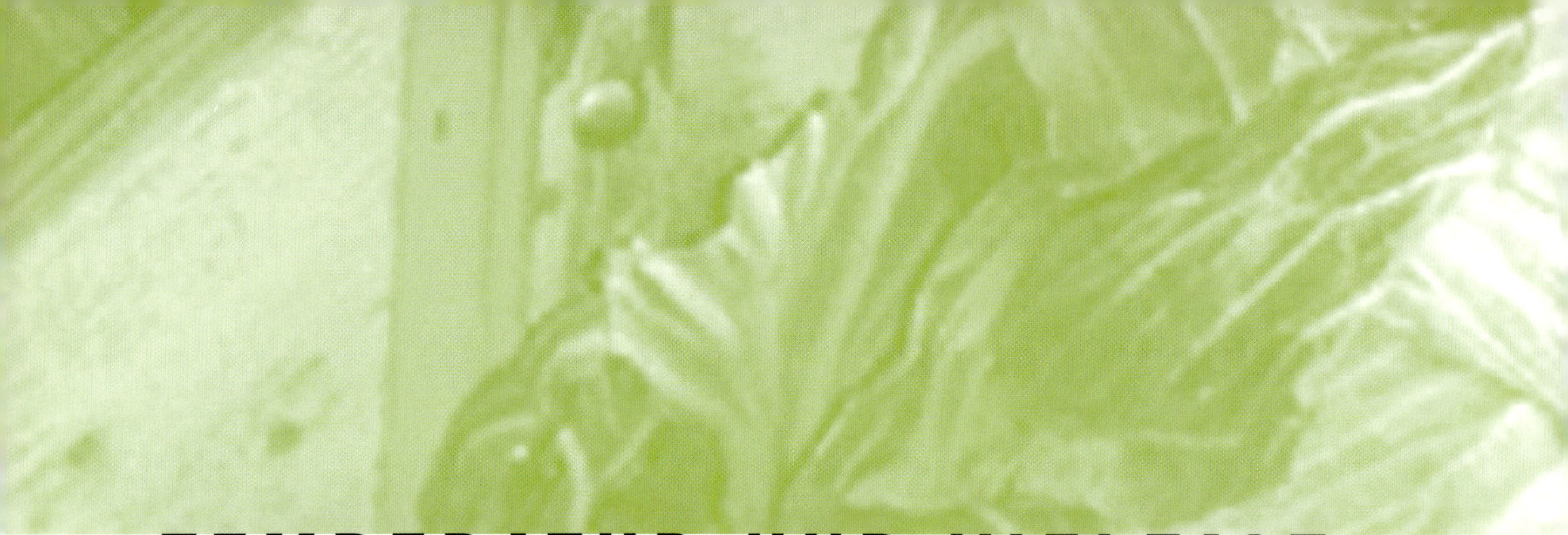

TEMPERATUR UND VIELFALT

SALAT À LA TASSILO

Salate sind für mich das Köstlichste vor dem Hauptgang, da sie die Magensäfte animieren. Der Salat muss nussig schmecken und knackig sein, und die Basis der Vinaigrette sollte Olivenöl und Rotweinessig sein – zum Beispiel aus Spanien, Italien, Frankreich oder Griechenland. Der Salat wird noch anregender, wenn Sie ein paar Tropfen frisch ausgepressten Knoblauchsaft in das Öl geben.

1 Den Salat putzen, waschen und trockenschleudern. Mundgerecht zerteilen oder schneiden. Das Salatblatt sollte höchstens so groß wie ein Rucolablatt sein.

2 Den gemischten Salat in eine große Schüssel geben. Die Tomaten waschen, putzen und in Scheiben schneiden. Olivenöl, Rotweinessig, Salz und Pfeffer nach Geschmack mischen und zusammen mit den Tomatenscheiben vorsichtig unter die Salatmischung heben.

3 Dazu passt ein kühler Prosecco, aber auch ein weißer oder roter Bordeaux.

1/2 Salatkopf Romana

1/4 Frisée

1 Lollo Rosso

1/2 Batavia

1 Bund Rucola

4–5 aromatische Tomaten

80 ml Olivenöl

40 ml Rotweinessig

Salz

schwarzer Pfeffer aus der Mühle

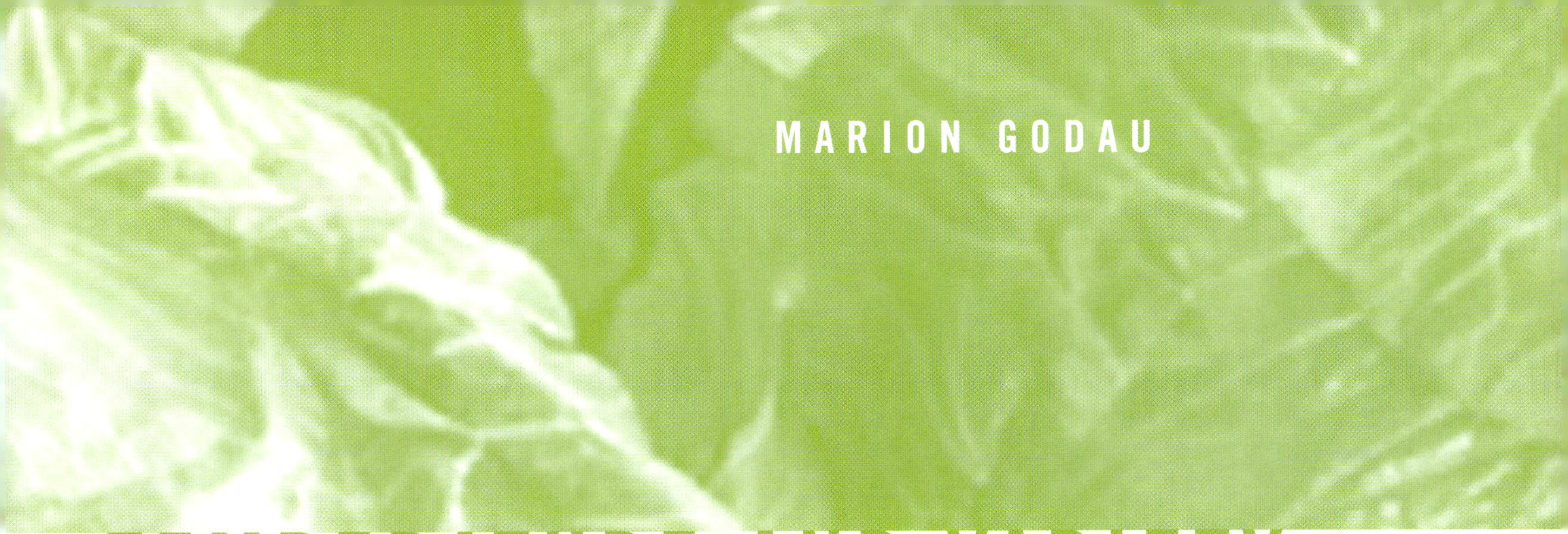

MARION GODAU

TEMPERATURE AND VARIETY

SALAD À LA TASSILO

For me, salads are the most delicious dish before the main course, as they stimulate the digestive juices. The salad must taste nutty and be crisp, and the basis of the vinaigrette should be olive oil and red wine vinegar – for example from Spain, Italy, France or Greece. The salad becomes even more stimulating if you add a few drops of freshly pressed garlic juice to the oil.

- *1/2 Romana lettuce*
- *1/4 Frisée lettuce*
- *1 Lollo Rosso lettuce*
- *1/2 Batavia lettuce*
- *1 bunch Rucola*
- *4–5 savoury tomatoes*
- *80 ml olive oil*
- *40 ml red wine vinegar*
- *Salt*
- *Milled black pepper*

1 *Clean the lettuce, wash and dry by spinning. Tear or cut into bite-sized pieces. The lettuce leaf should be at most as large as a rucola leaf.*

2 *Put the mixed salad into a large bowl. Wash and clean the tomatoes and cut into slices. Mix the olive oil, red wine vinegar, salt and pepper according to taste and, together with the tomato slices, carefully fold into the salad mixture.*

3 *A cool Prosecco goes well with this dish, alternatively a white or red Bordeaux.*

Als der Brite James Dewar um 1892 die schwierige Aufgabe meisterte, ein doppelwandiges Isoliergefäß zu konstruieren, dachte er dabei weder an heißen Tee noch an kühle Limonade, sondern an Laborgase. Jene flüchtigen Elemente nehmen bei bestimmten Temperaturen einen flüssigen Aggregatzustand an und lassen sich dann in Flaschen gefüllt lagern. Um deren Innentemperatur so lange wie möglich konstant zu halten, nutzte Dewar die isolierenden Eigenschaften des Vakuums, indem er dem Zwischenraum zweier Glashüllen die Luft entzog. Eine zusätzlich dünn aufgebrachte Silberschicht auf der inneren Flaschenwandung verstärkte den Isoliereffekt. Uneigennützig stellte der Physiker seine Erfindung als Geschenk für die Wissenschaft zur Verfügung. Das „Dewar-Gefäß", wie es bald hieß, wurde fortan dazu genutzt, flüssigen Stickstoff zu lagern, Impfstoffe aufzubewahren oder tropische Fische zu transportieren. James Dewar ahnte damals nicht, dass seine Labor-Flaschen rund zehn Jahre später auch Haushalte erobern würden. Der Berliner Unternehmer Reinhard Burger, in dessen Betrieb Dewar seine Vakuumflaschen produzieren ließ, erkannte die vielfältigen kommerziellen Möglichkeiten der Erfindung zuerst. Burger ergänzte Dewars Flasche mit einer Hülle aus Metallblech und verringerte ihre Größe etwas. Seine kleinen Veränderungen erweiterten die Haltbarkeit der zerbrechlichen Glaswände enorm. Als Thermosflaschen oder -kannen kamen sie 1903 auf den Markt. Ihre praktischen Vorzüge überzeugten sofort, und bald konnten Leser der immer beliebter werdenden Illustrierten verfolgen, wie sich sowohl Robert E. Peary am Nordpol als auch Ernest Henry Shakleton am Südpol mit dampfenden Heißgetränken aus Thermosflaschen von den Strapazen ihrer Expeditionen erholten. Auch die Gebrüder Wright und Graf Zeppelin, Pioniere bei der Eroberung der Lüfte, einte der Griff zur Thermoskanne. Da ist es kein Wunder, dass das Behältnis schnell in Wohnungen wie Werkshallen gesichtet wurde. An Dewars Isolier-Prinzip beziehungsweise Burgers Thermoskanne hat sich bis heute nichts Grundlegendes geändert. Besonders Globetrotter und die Hotellerie schätzen inzwischen jene weiterentwickelten Isolierkannen-Varianten, deren Vakuum-Ummantelung nicht aus Glas-, sondern robusten Metallwänden besteht und daher so gut wie unverwüstlich ist.

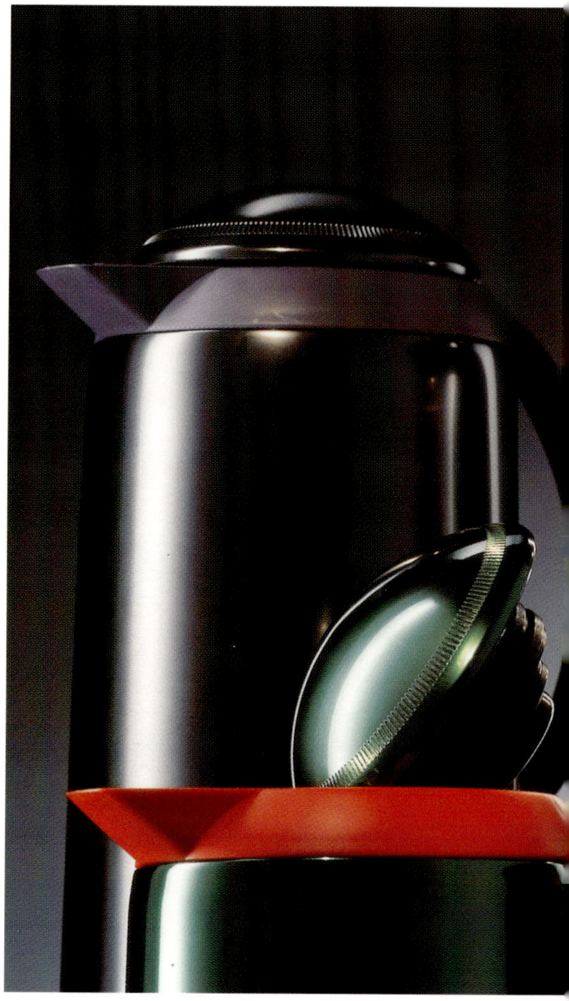

When the Briton James Dewar mastered the difficult exercise of constructing a twin-walled insulating vessel in around 1892, he was thinking neither of hot tea nor cool lemonade but laboratory gases. Those volatile elements become liquid at certain temperatures and can then be stored in flasks. In order to keep their inner temperature constant for as long as possible, Dewar used the insulating properties of the vacuum through removing the air from the space between two glass sleeves. An additional thinly applied layer of silver on the inner wall of the flask strengthened the effect of insulation. The physicist unselfishly donated his discovery to science. The "Dewar flask", as it soon became known, was from this time on used for storing liquid nitrogen, keeping vaccination materials or transporting tropical fish. At the time James Dewar had no idea that his laboratory flasks would also conquer households around ten years later. The Berlin entrepreneur Reinhard Burger, whose company manufactured Dewar's vacuum flasks, was the first to recognize the diverse commercial possibilities of the invention. Burger added a thin metal shell to Dewar's flask and made it slightly smaller. His small changes extended the durability of the fragile glass walls enormously. In 1903 they came on the market as thermos flasks or pots. Their practical advantages were immediately apparent and readers of the increasingly popular magazines could follow how Robert E Peary at the North Pole as well as Ernest Henry Shakleton at the South Pole recovered from the strains of their expeditions with steaming hot drinks from thermos flasks. The Wright brothers and Count Zeppelin, pioneers of the conquest of the air, were also united by the use of the thermos flask. It is no wonder then that this receptacle was quickly to be seen in the home and workplace. Today there is no fundamental deviation from Dewer's insulation principle, or Burger's thermos flask. In the meantime, particularly globetrotters and the hotel business have come to value those further developed insulating pot variations. The vacuum overcoats are no longer made from glass but have robust metal walls and are therefore as good as indestructible.

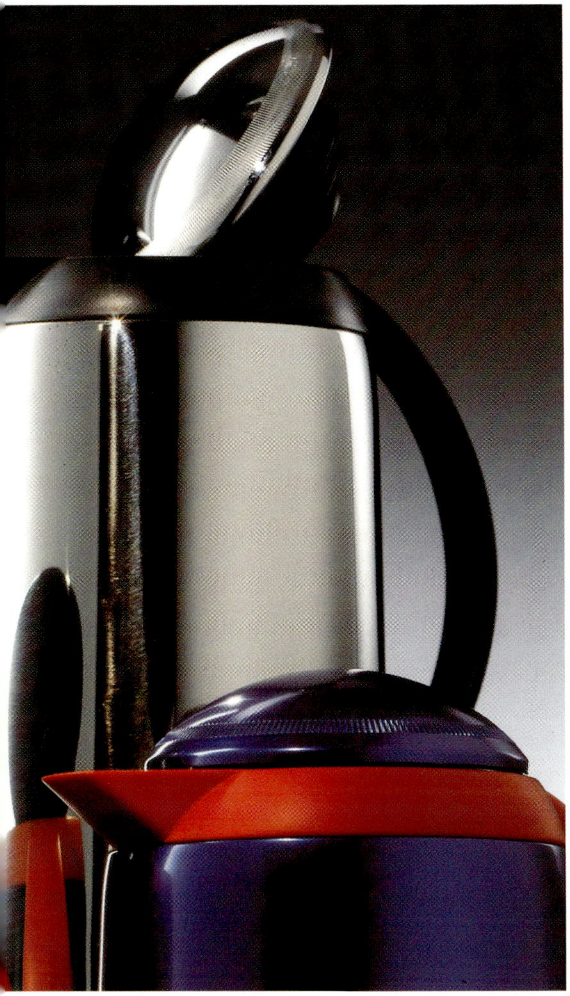

„Lipstick" gibt es in mehreren Ausführungen und Farben: Aluminium farbig eloxiert, Edelstahl hochglanzpoliert. 1988 für Alfi.

"Lipstick" is available in several variations and colours: coloured aluminum, high-gloss stainless steel. 1988 for Alfi.

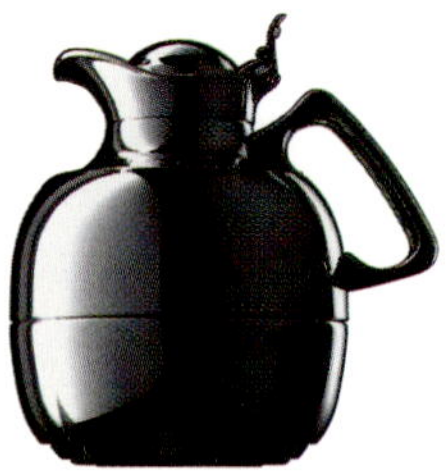

EIN GUTES TEAM

Nur wenig jünger als die Erfindung der Thermoskanne ist die „Aluminiumfabrik Fischbach", heute als Alfi bekannt. 1914 von den Eheleuten Carl und Sophie Zitzmann gegründet, gelang Alfi schon vier Jahre später mit der Isolier-Kaffeekanne „Juwel" ein Evergreen, der noch heute in sechs verschiedenen Größen im Programm ist. Heute führt Jörg Dümmig-Zitzmann die Alfi-Tradition hochwertiger Designentwürfe fort. 1984 begann das Unternehmen mit dem Dänen Ole Palsby die gezielte Zusammenarbeit mit namhaften Gestaltern. Tassilo von Grolmans Arbeit für Deutschlands erfolgreichsten Isolierkannen-Hersteller begann wenige Jahre später. Als Erstes unterzog er Alfi-Klassiker „Juwel" einem Facelifting, bei dem der Ausguss unverändert blieb, Verschluss und Stöpsel aber ergonomischer wurden. Später kreierte er eine dazu passende „Juwel"-Teekanne. Heute können Alfi und von Grolman auf eine ganze Sammlung von gemeinsamen Isolierprodukten zurückblicken. Das kommerziell erfolgreichste Produkt dieser Zusammenarbeit heißt „Achat". Von vornherein wurde die Produktfamilie in verschiedenen Formvarianten konzipiert. Designer von Grolman sollte „Juwels" Traditionsverschluss übernehmen, dabei aber einen modernen Kannenkorpus für unterschiedliche Zielgruppen finden und den Griff an nur einem Punkt befestigen. Als verbindende Elemente fand er abgeschrägte, strenge Griffe und eine gerade Kegelform, die er auf dreierlei Weise variierte: geriffelt, glatt und facettiert. Alte Emailkannen dienten als Ideenlieferanten. Die glatte Version für Formpuristen „ist sicherlich die schlichteste und schönste, aber die unpraktischste, weil Sie jeden Flecken, Fingerabdruck et cetera auf dem Körper sehen", findet von Grolman und erklärt, störende Flecken würden bei der geriffelten Version weitgehend vermieden, indem sie durch ihre Gliederung nur kleine Flächen widerspiegeln. Fingerabdrücke fielen einfach weniger auf. Zudem gäben Riffelungen der Kanne eine moderne und gleichzeitig sympathische Wirkung. „Achat"-Version Nummer drei ist für eine eher konservativ orientierte Zielgruppe gedacht und vereint die praktischen Vorzüge einer gegliederten Fläche mit einem traditionellen Aussehen. Zu ihrem Handgriff kamen alle drei Modelle eher zufällig. Von Grolman hatte, um einen Henkel am Kannenmodell zu simulieren, einen harten Schlauch angesetzt und ihn schräg abgeschnitten. Erfreut stellte er fest, dass die Behelfsform optimal zu den verschiedenen Formvarianten passte und beließ es dabei.

Linke Seite
Isolierkanne „Tee Juwel" für Alfi, 1987.

Rechte Seite
Drei verschiedene Varianten – drei verschiedene Wirkungen: Isolierkanne „Achat" für Alfi, 1990.

Left-hand side
Thermal pot "Tee Juwel" for Alfi, 1987.

Right-hand side
Three different variations – three different effects: Thermal pot "Achat" for Alfi, 1990.

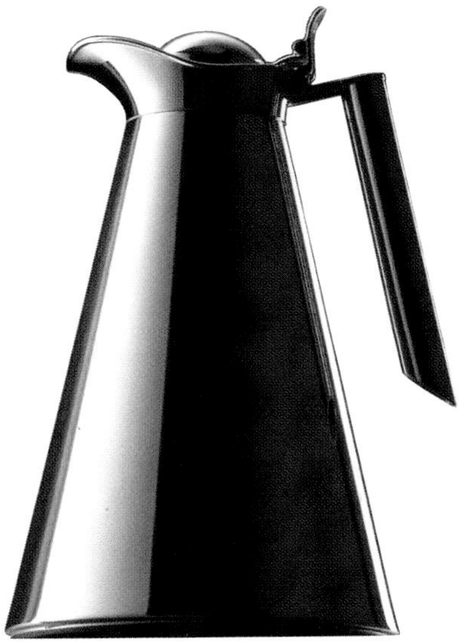

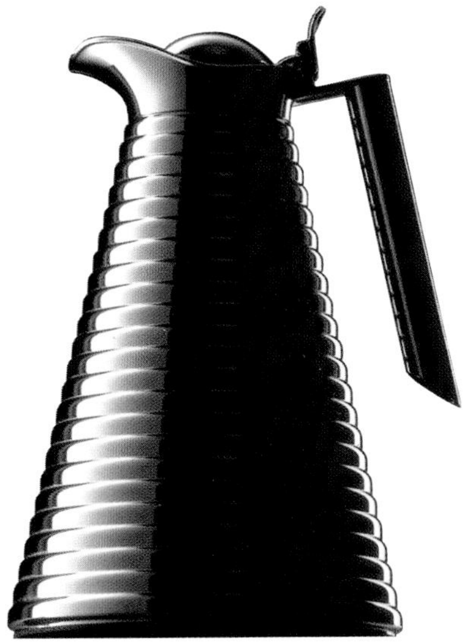

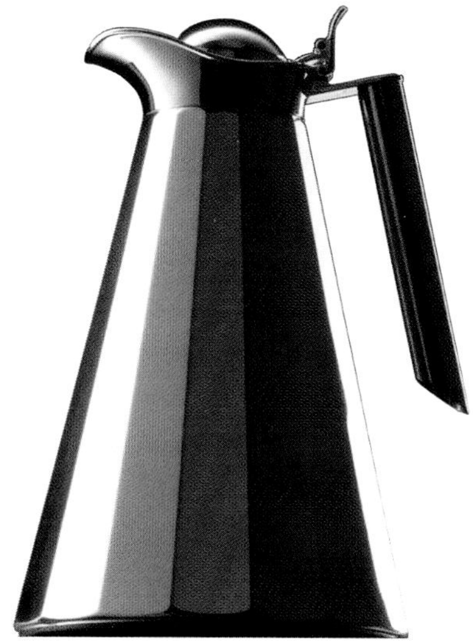

A GOOD TEAM

Just a little younger than the invention of the thermos flask, the "Aluminiumfabrik Fischbach" is today known as Alfi. Founded by husband and wife Carl and Sophie Zitzmann in 1914, Alfi enjoyed success only four years later with the insulating coffee pot "Juwel", an evergreen which is still in the range today in six different sizes. Today Jörg Dümmig-Zitzmann continues the Alfi tradition of high value designs. In 1984, the company began its focused cooperation with renowned designers with the Dane Ole Palsby. Tassilo von Grolman's work for Germany's most successful manufacturer of insulating pots began a few years later. First of all he subjected the Alfi classic "Juwel" to a face-lift, in which the spout remained unchanged but the lid and stopper became more ergonomic. Later he created a matching "Juwel" teapot. Today Alfi and von Grolman can look back on a whole collection of insulating products from their collaboration. The most commercially successful result of this collaboration is called "Achat". From the outset, the product family was conceived as having variations in shape. Designer von Grolman was supposed to take over the traditional seal of "Juwel" but with it find a modern body for the pot for various target groups, and secure the handle at just one point. As a connecting element he found sloping, serene handles and a straight cone shape, which he varied in three different ways: rippled, smooth and facetted. Old enamel pots served as deliverers of ideas. Von Grolman thinks that the smooth version for shape purists "is certainly the most simple and the most beautiful, but the most impractical because you see every stain, fingerprint et cetera on the body". He explains that disturbing stains would be avoided to a great extent with the rippled version as it only reflects small stains because of its structure. Fingerprints were simply less conspicuous. Moreover the ripples would give the pot a modern and simultaneously pleasant effect. "Achat" version number three is intended for a more conservatively orientated target group and unites the practical advantages of a structured surface with a traditional look. All three models got their handles rather by chance. In order to simulate the handle on the pot model, von Grolman had attached a hard piece of hose and cut it down at a slant. He was pleased to find out that the temporary form suited the different variations in shape very well and left it as it was.

Isolierkanne „Big Mama“ fasst 1,7 Liter,
Hersteller Alfi, 1993.

*Thermal pot “Big Mama” holds 1.7 litres,
manufacturer Alfi, 1993.*

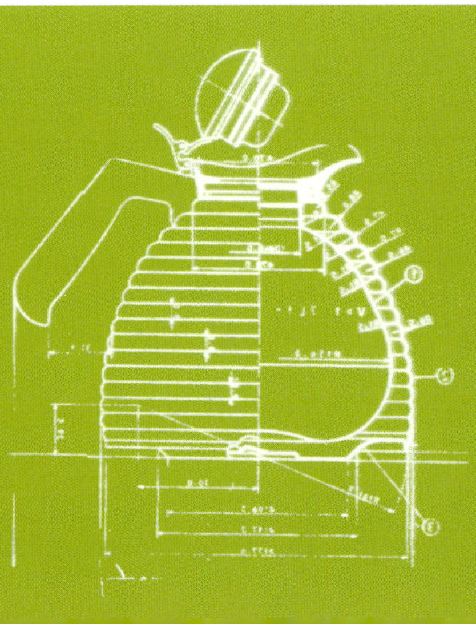

ZU SCHÖN, UM ERFOLGREICH ZU SEIN

Von all seinen Isolierkannen-Entwürfen verkauft sich „Achat“ am besten, von Grolmans Lieblingskanne jedoch ist „Miss Ellie“. „Meine schönste Kanne“, sagt der Designer und ist besonders stolz darauf, dass der französische Luxusartikelkonzern Hermès sie in einer Edelversion mit Versilberung und ledernem Griffüberzug weltweit in seinen Boutiquen vertreibt – zu saftigen Preisen versteht sich. Dabei war „Miss Ellies“ elliptische Form ursprünglich für den Hotelbetrieb gedacht. Dort ist das Decken der Hoteltische eine zeit- wie laufaufwendige Angelegenheit, weshalb das Personal so viel Geschirr wie möglich auf einmal trägt. Zwar war von Grolmans ovale Lösung sehr aufwendig und schwierig zu fertigen, durch ihr robustes Edelstahl-Innenleben überstand sie jedoch auch den ruppigsten Umgang. „Miss Ellies“ zierliche Ellipsen-Eleganz war nicht nur praktisch, sondern gefiel Personal wie Gästen. Letzteren leider zu sehr, denn „Miss Ellie“ wurde so oft gestohlen, dass selbst Luxushotels wie das Ritz Carlton bedauernd auf die schöne Form verzichteten. Ein ähnliches Schicksal ereilte die Gastronomiekanne „Big Mama“, die der Designer für den US-amerikanischen Markt erdachte. „Big Mama“, deren Name auf ihr großes Fassungsvermögen von 1,7 Litern anspielt, sollte die in den USA üblichen Glaskannen ersetzen. Auf Wärmeplatten abgestellt blieb deren Inhalt zwar einigermaßen heiß, ließ geschmacklich aber zu wünschen übrig. Außerdem sahen die Glaskannen schnell fleckig aus und zerbrachen des öfteren. „Big Mama“ schuf Abhilfe, denn ihr Kannenkörper bestand innen wie außen aus unverwüstlichem Edelstahl. Den technischen Vorsprung gegenüber anderen Gastronomiekannen ergänzte von Grolman mit einem für unterschiedlichste Hände und schnelle Handhabung ausgeformten Griff und einer Rillenstruktur, die Kaffeeflecken optisch in den Hintergrund treten ließ. Trotz ihres großen Fassungsvermögens wirkt „Big Mamas“ bauchige Gestalt dabei relativ klein – ein Spitzenentwurf, für den Alfi hohe Herstellungskosten in Kauf nahm. Vorbildlich gestaltet, haltbar, generationenlange Lebensdauer – gerade diese Eigenschaften verhinderten „Big Mamas“ großflächige Verbreitung. Leider ist sie wie „Miss Ellie“ ein unfreiwilliges „Give-away“.

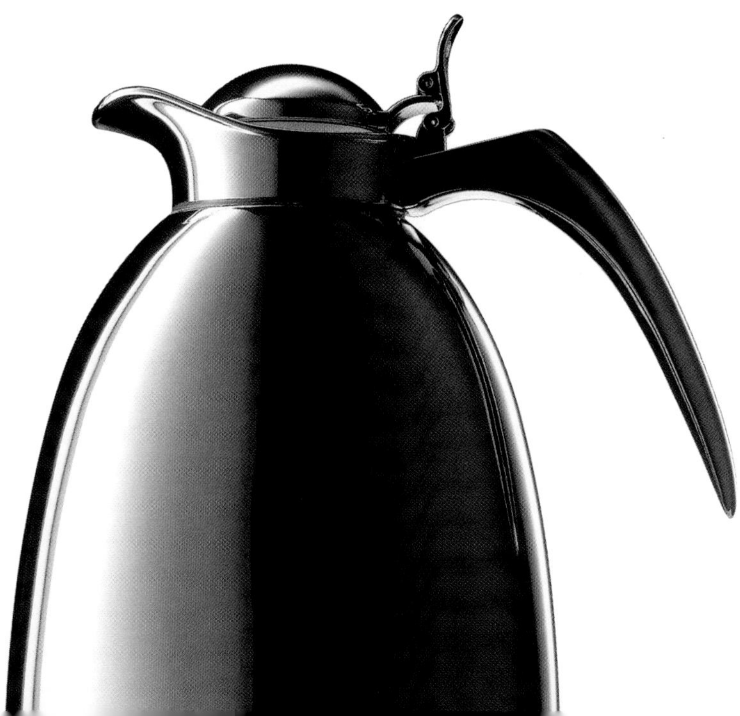

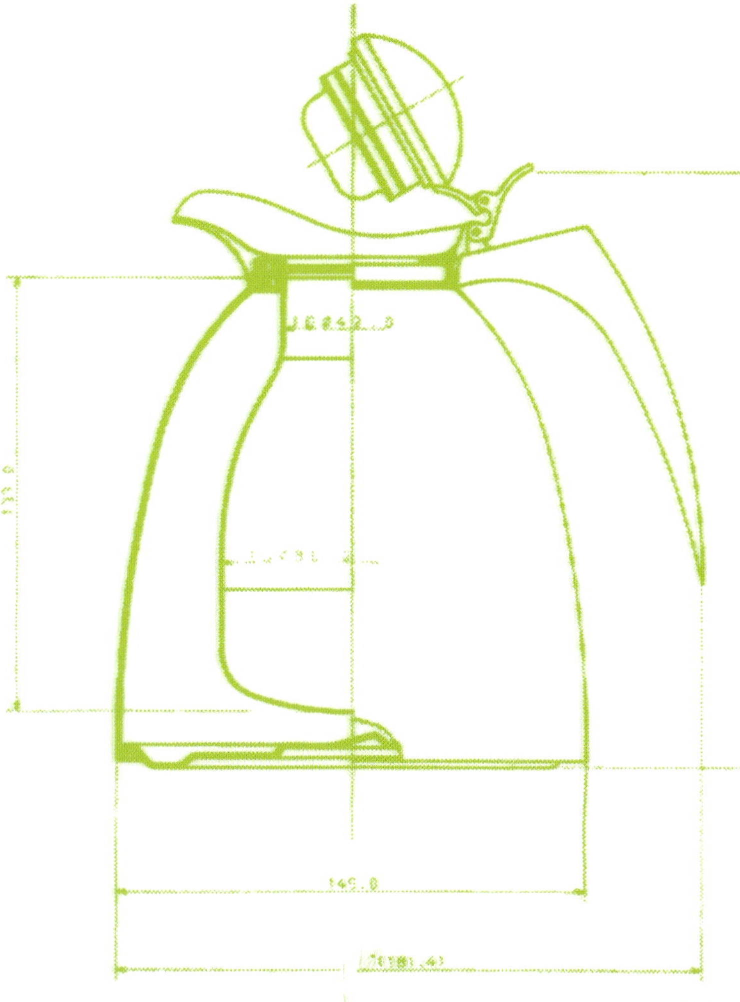

Von Grolman gab „Miss Ellie“ ihren Namen wegen des elliptischen Grundrisses. Bei Hersteller Alfi heißt die Isolierkanne für den professionellen Bereich „HotelDesign“. Auch wenn es scheint, als seien es zwei Kannen, so ist es doch „Miss Ellie“ aus verschiedenen Perspektiven, 1993.

Von Grolman gave "Miss Ellie" that name because of its elliptical shape. The manufacturer Alfi calls the thermal pot for professional use "HotelDesign". Although it seems to be two pots, it is only "Miss Ellie" from different perspectives, 1993.

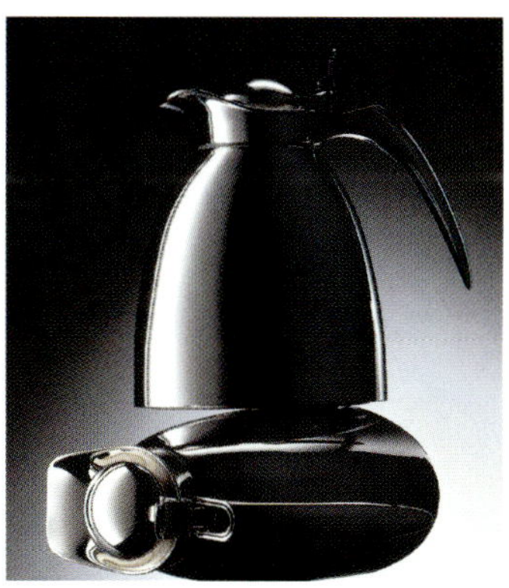

TOO BEAUTIFUL TO BE SUCCESSFUL

Of all his insulating pot designs, "Achat" is the best seller. Von Grolman's favourite pot however is "Miss Ellie". "My most beautiful pot" says the designer, and is particularly proud that the French luxury goods company Hermès features it worldwide in its shops in a silver-plated special edition with a leather handle cover – of course at hefty prices. Yet the elliptical shape of "Miss Ellie" was originally intended for the hotel business. There the laying of hotel tables is a time and energy consuming business, where the staff carries as much tableware as possible in one go. Although von Grolman's oval solution was actually very costly and difficult to make, it survived even the roughest treatment due to its robust stainless steel interior. The dainty elliptical elegance of "Miss Ellie" was not only practical but also appealed to staff as much as to guests. The latter a little too much, as "Miss Ellie" was stolen so often that even luxury hotels like the Ritz Carlton regretfully had to do without the beautiful shape. A similar fate befell the gastronomic pot "Big Mama", which the designer had intended for the US market. "Big Mama", whose name is a play on its large capacity of 1.7 litres, was supposed to replace the customary glass pots in the USA. Their contents certainly remained reasonably warm when put on a hotplate, but the taste did leave something to be desired. In addition to this they quickly looked stained and often broke. "Big Mama" provided a remedy as the body of the pot was made inside and out from indestructible stainless steel. Von Grolman extends the technical advantage over other gastronomic pots with a handle shaped for all types of hands and for quick handling, and a grooved structure that optically pushes coffee stains into the background. In spite of its large capacity, the bulbous shape of "Big Mama" makes it seem relatively small – a top design, for which Alfi accepted high manufacturing costs. Exemplary design, durable, lasting for generations – precisely these properties hindered the extensive distribution of "Big Mama". Like "Miss Ellie", it is unfortunately an unintentional "give away".

1001 MÖGLICHKEIT

Ganz anders dagegen Alfi-Kanne „Change". „Change" macht hohe Umsätze im Werbemittelbereich und ist, wie Alfi-Geschäftsführer Jörg Dümmig-Zitzmann sagt, „ein Renner". Er selbst war Anfang der neunziger Jahre auf die Idee gekommen, eine Kanne zu vertreiben, die ihr Gesicht immer wieder verändern kann und damit als Werbemittel oder als Kanne für besondere Gelegenheiten vielfache Variationsmöglichkeiten bietet. Das hieß für Dümmig-Zitzmann, dass die Wandelkanne rundum bedruckbar sein sollte. Tassilo von Grolmans Lösung setzte die Vorgaben nicht nur um, sondern überschwemmte den Alfi-Geschäftsführer förmlich mit Gestaltungsmöglichkeiten. „Change" bekam einen bedruckbaren langgestreckten Kunststoffkorpus mit zwei vertikal verlaufenden Streben. Wechselrahmen und Thermoskörper brauchen dann nur noch zusammengeschraubt werden. Ein großer Ausguss sorgt dafür, dass die Mini-Litfasssäule auch ihre Pflichten als Kanne erfüllt. Ob als Corporate-Design-Utensil oder als passendes Accessoire für Hochzeiten und andere Festtage, Handel wie Laie können in Minutenschnelle immer wieder eine andere Kanne kreieren – was sie, wie von Grolman bedauert, in der Praxis selten tun.

1001 POSSIBILITIES

Alfi pot "Change" is completely different. "Change" achieves high sales in the merchandising business and is, as Alfi's managing director, Jörg Dümmig-Zitzmann, says "a hit". In the early nineties, he himself hit upon this idea of selling a pot which can always change its face, thereby offering many possibilities for variation as a means of advertising or as a pot for special occasions. For Dümmig-Zitzmann, this meant that the changeable pot should be printable all the way round. Not only did Tassilo von Grolman's solution meet the requirements, but positively flooded Alfi's managing director with design possibilities. "Change" gained a printable stretched plastic body with two vertically running struts. Clip-on frames and thermal bodies then just need screwing together. A large spout ensures that the mini advertising column also fulfils its duties as a pot. Whether as corporate design utensil or as matching accessory for weddings and other special days, retailers and laymen can create a different pot again and again in minutes – something that, as von Grolman regrets, they seldom do in practice.

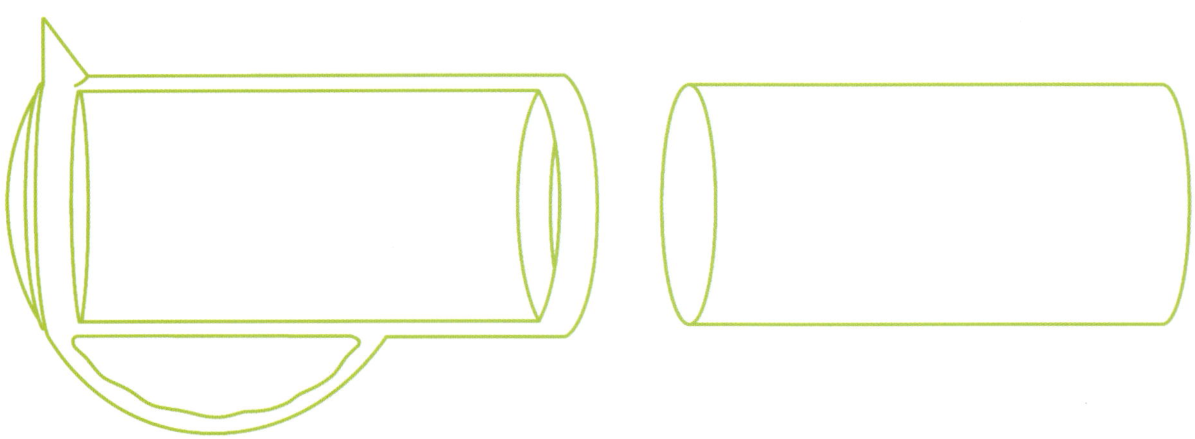

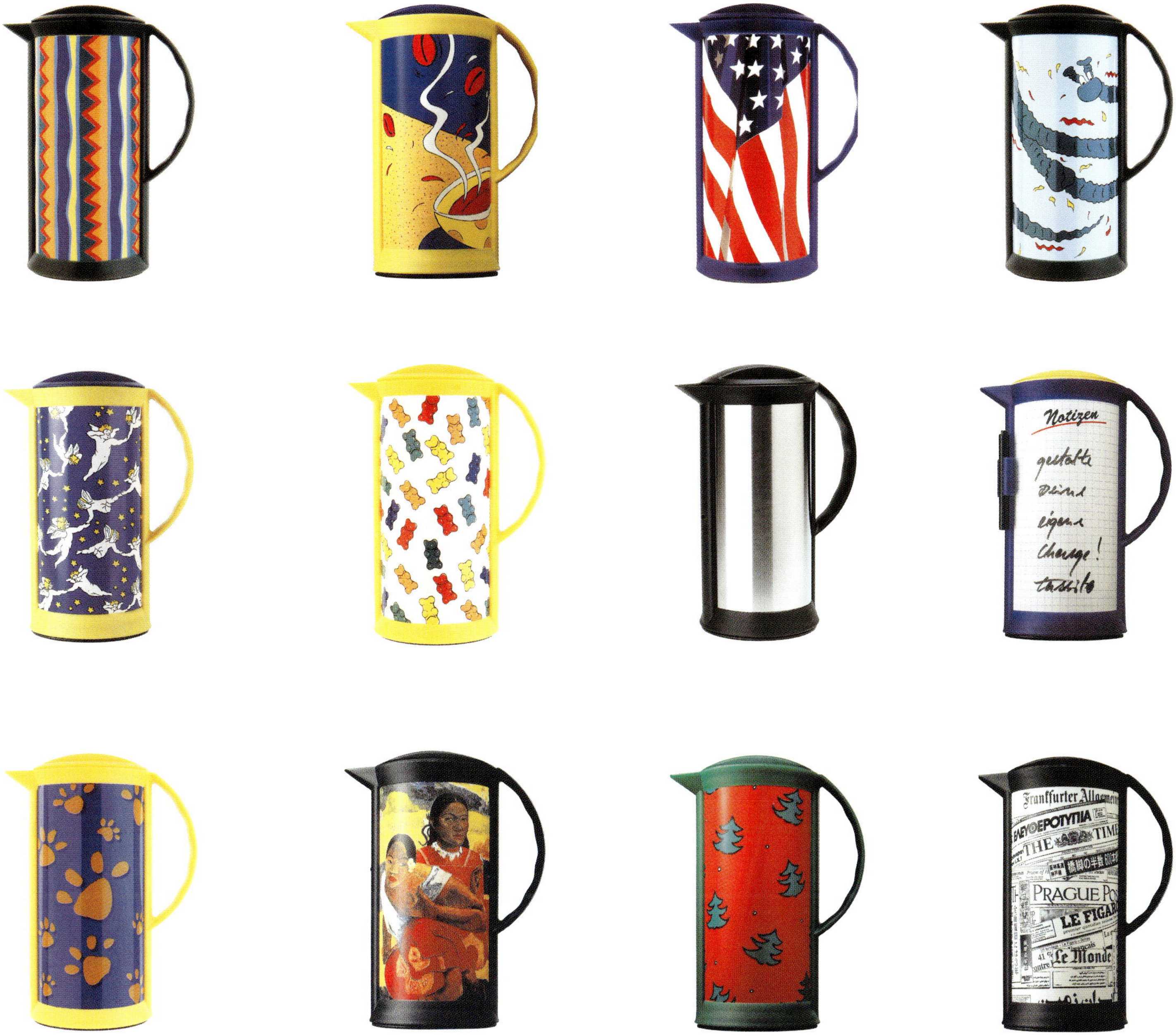

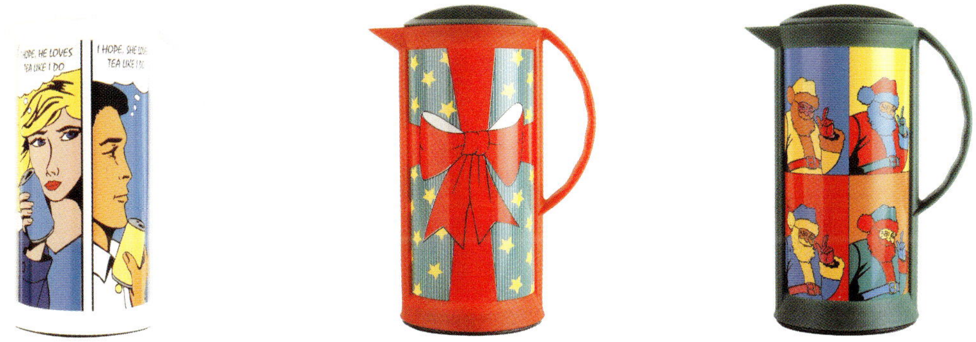

Einige der unzähligen Möglichkeiten der Isolierkanne „Change“ für Alfi, 1994. Von Grolmans Produkte für Alfi sind langfristig im Programm. „Change“ ist besonders erfolgreich.

Some of the countless possibilities of the thermal pot "Change" for Alfi, 1994. Von Grolman's products for Alfi are permanently in the program. "Change" is especially successful.

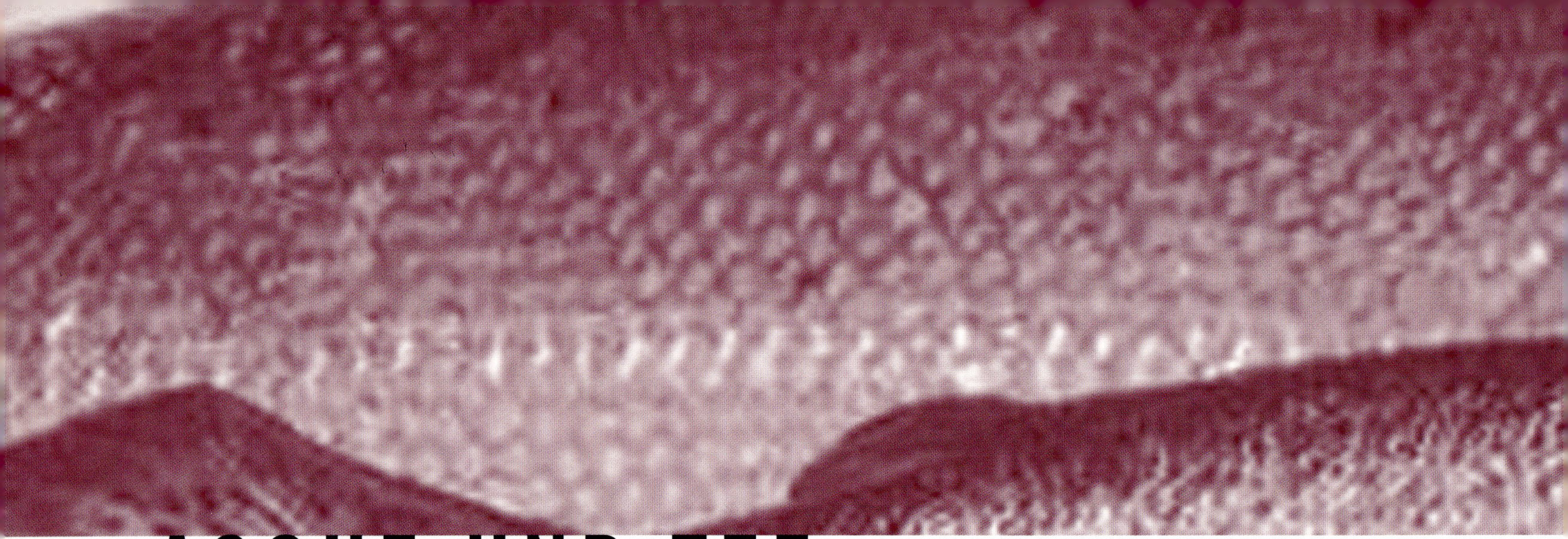

ASCHE UND TEE

LACHS MIT FETTUCCINE

1 Die Zwiebel und den Knoblauch schälen und die Petersilie waschen. Zwiebel, Knoblauch und Petersilie klein hacken und jeweils in kleine Schüsseln geben.

2 Den Lachs enthäuten und das Lachsfiletstück in kleine ungefähr 1 cm große Würfel schneiden.

3 In der Pfanne Olivenöl erhitzen, die Butter, die Zwiebel und den Knoblauch zugeben. Wenn die Zwiebel glasig ist, den Lachs von allen Seiten kurz anbraten und mit dem Wein ablöschen.

4 Fischfond, Crème fraîche und Paprika zufügen und mit Salz abschmecken. Das Ganze nicht länger als 3 Minuten köcheln lassen. Mit dem rosa Pfeffer würzen.

5 Die Fettuccine in reichlich Salzwasser 2 Minuten kochen, in ein Sieb abgießen, abtropfen lassen und auf Tellern anrichten. Den Lachs darüber geben und mit Petersilie bestreuen.

1 Zwiebel

1 Knoblauchzehe

frische Petersilie

1/2 frische Lachsseite

1 EL Olivenöl

50 g Butter

100 ml trockener Weißwein

1 Glas Fischfond

125 g Crème fraîche

1 Msp. edelsüßer Paprika

Salz

1 EL rosa Pfeffer

(im Mörser gestoßen)

500 g frische rosa Fettuccine

(im Kühlregal)

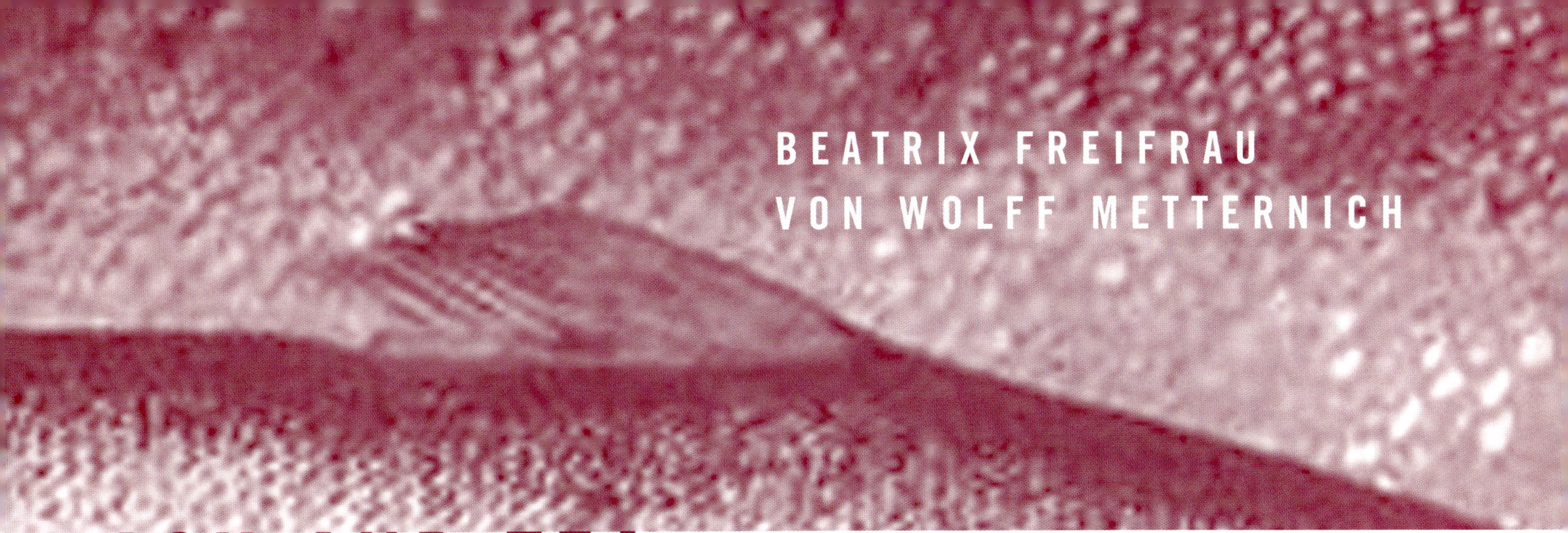

BEATRIX FREIFRAU VON WOLFF METTERNICH

ASH AND TEA

1 onion

1 clove of garlic

Fresh parsley

1/2 side of fresh salmon

1 tbsp olive oil

50 g butter

100 ml dry white wine

1 glass fish stock

125 g crème fraîche

1 pinch mild paprika

Salt

1 tbsp red peppercorns

(crushed with a pestle and mortar)

500 g fresh red fettuccine

(from the refrigerator cabinet)

SALMON WITH FETTUCCINE

1 *Peel the onion and garlic and wash the parsley. Chop the onion, garlic and parsley finely and put each into a small bowl.*

2 *Skin the salmon and cut the salmon fillet into cubes of around 1 cm.*

3 *Heat the olive oil in the pan, and add the butter, onion and garlic. When the onion is transparent, quickly brown the salmon on all sides and pour in the wine.*

4 *Add the fish stock, crème fraîche and paprika and season with salt. Allow the whole mixture to simmer for no longer than 3 minutes. Season with the crushed red peppercorns.*

5 *Boil the fettuccine in plenty of salted water for 2 minutes. Pour the fettuccine into a sieve, allow to drain and serve out onto plates. Place the salmon on top and sprinkle with parsley.*

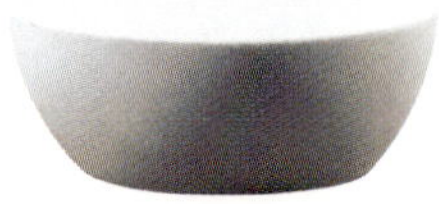

Es gibt wohl kaum einen namhaften Entwerfer oder Designer, der nicht irgendwann in seinem kreativen Leben der Verführung erlegen ist, sich auch einmal auf den Werkstoff Porzellan einzulassen. Eine Einlassung ist dies in jedem Fall: Erst kommt die Verführung durch das makellos reine Weiß des Werkstoffs, das durch die Glasur in seinem unverwechselbaren Schmelz erscheint. Dann beginnt die Auseinandersetzung mit der Erkenntnis, wie weit der Weg vom Entwurf bis zur Vollendung eines Porzellanobjekts ist. Als Tassilo von Grolman sich mit dem Auftrag befasste, eine Teedose für die Ausformung in Porzellan zu gestalten, stand für ihn allerdings der Inhalt des zu gestaltenden Objekts im Vordergrund. Er konzentrierte sich auf die Essenz, die duftenden Teeblätter. Hier war Tassilo von Grolman in seinem Element, denn wenn er von der Zubereitung des Tees, von den Finessen und Ritualen des Tee-Zeremoniells spricht, spürt man, worum es ihm in seinen Arbeiten nicht unwesentlich geht: um Stil, um die Hervorkehrung des Wesentlichen, sei es in der Durchführung eines uralten Rituals oder in der Gestaltung eines Objekts.

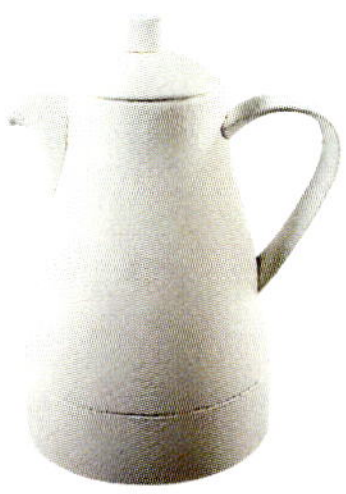

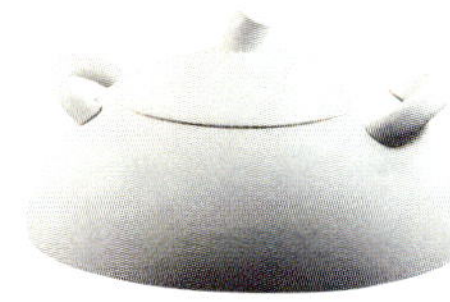

There is probably hardly a renowned designer who has not succumbed to temptation at some point in his creative life and become involved at least once in working with porcelain. This is a case of letting oneself in for it: the temptation comes first through the flawless pure whiteness of the material, which appears through the enamel with its unmistakable lustre. Then the analysis begins with the realisation of how far it is from design to completion of a porcelain product. When Tassilo von Grolman was dealing with the task of creating a tea caddy form in porcelain, the contents of the object to be designed were his focus. He concentrated on the essence, the fragrant tea leaves. Tassilo von Grolman was in his element here – when he is talking about tea preparation, about the finesse and ritual of the tea ceremony, it is possible to sense what his work is not insignificantly all about: style, putting the focus on what is important, whether it is carrying out an ancient ritual or the design of an object.

Kaffee- und Teeservice,
Studie für Arzberg, 1992.

*Coffee and tea service,
study for Arzberg, 1992.*

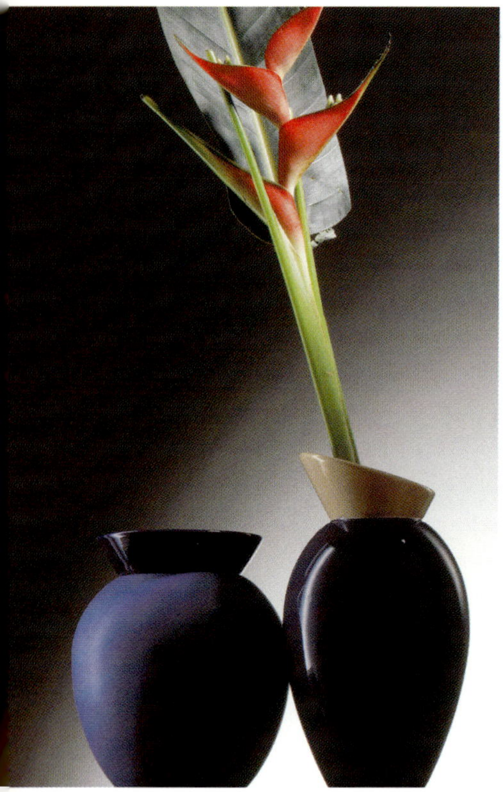

KOSTBARER WERKSTOFF PORZELLAN

Sowohl der Tradition des Teetrinkens als auch dem Porzellan haften bis heute etwas Geheimnisvolles an. Beide stammen aus China und waren dort schon Jahrtausende bekannt. In Europa feierte zu Beginn des 18. Jahrhunderts Tee als eines der großen Modegetränke seinen Durchbruch. Sicher ist es kein Zufall, dass die Erfindung des europäischen Porzellans in Meißen zeitlich damit zusammenfiel. Wie wohl kein zweiter Designer hat Tassilo von Grolman über die Tradition von Tee nachgedacht und mit seinen Kannenerfindungen die Teezubereitung revolutioniert. Kein Wunder also, dass der Designer von dem alten, 1706 gegründeten englischen Teehaus Twinings beauftragt wurde, zu dessen dreihundertjährigem Jubiläum ein ganz besonderes Objekt aus Porzellan zu gestalten: eine Teedose. Dieses Behältnis ist auch in Europa ein traditionsreicher Gegenstand, der meistens aus Silber oder Porzellan hergestellt wurde. Im 18. Jahrhundert gehörten Teedosen aus Porzellan zu jedem Teeservice. Mit den selben Motiven bemalt wie die übrigen Teile des Services, waren die Dosen relativ klein und fassten nur eine geringe Menge von Teeblättern. Sie reichte gerade für eine kleine Teegesellschaft, die Teemenge war für sechs Tassen berechnet. Zunächst in Meißen entwickelt, waren diese kleinen Porzellanbehältnisse während der Zeit des Rokoko gebräuchlich. Ihre hochrechteckige Form mit scharf abgesetzten Schultern wurde mit einem kleinen runden, teelöffelgroßen Deckel verschlossen. Er diente auch als Maß für die benötigte Teemenge in der Kanne. Gegen Ende des 18. Jahrhunderts zeigte sich die neue Stilform des Klassizismus auch in der Gestaltung der Teeservice. Teedosen wurden nun zylindrisch rund oder oval. Die scharf abgesetzten Schultern, kleinen Öffnungen und Deckel blieben, waren aber im Unterschied zu früher nicht mehr mit verspielten Blumen als Knauf, sondern mit klassischen Akanthusblüten oder Pinienzapfen versehen. Im 19. Jahrhundert verselbstständigte sich die Teedose zum Einzelgegenstand. Sie war nicht mehr zwingend Bestandteil eines bestimmten Services und trat auch in individueller Form auf. In England entwickelten sich die Behältnisse für Teeblätter als „Tea-Caddy“ immer mehr zum beliebten Geschenkartikel. Sie waren meist in Silber und in polygonaler Form oder als Schatulle gestaltet.

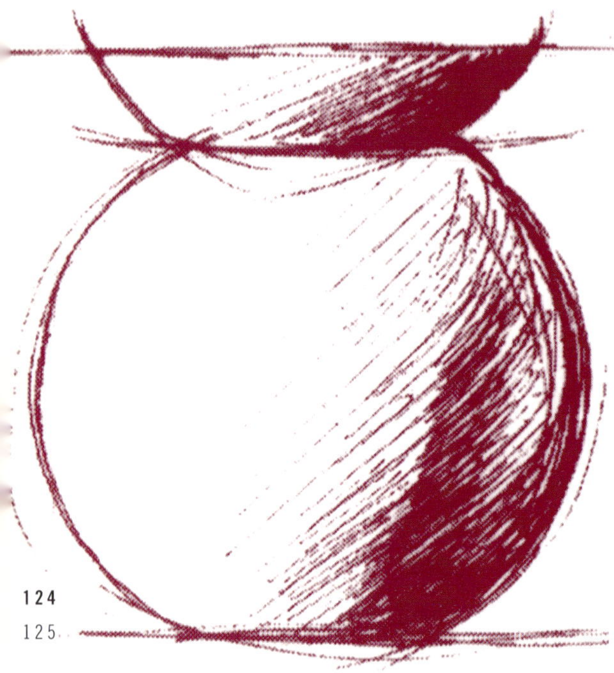

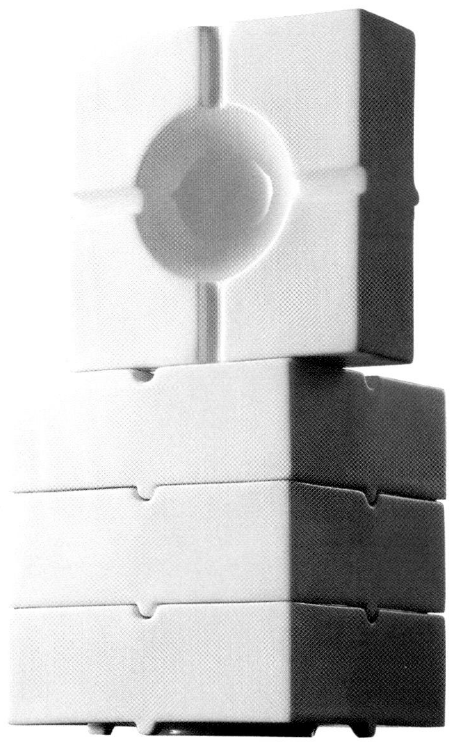

PORCELAIN – A VALUABLE MATERIAL

Up to the present day there is something mysterious about both the tradition of tea drinking and about porcelain. Both originate from China and have been known there for thousands of years. In Europe, tea made its breakthrough as a great fashionable drink at the beginning of the eighteenth century. It is certainly no coincidence that the invention of European porcelain in Meissen occurred around the same time. Tassilo von Grolman has reflected on the tradition of tea probably as no other designer and has revolutionized tea preparation with his teapot inventions. It is also no wonder that the designer was asked by Twinings, an English tea company founded in 1706, to design a very special object in porcelain for their three hundredth anniversary: a tea caddy. This receptacle also has a long tradition in Europe, and was mostly made from silver or porcelain. In the eighteenth century, every tea set had a porcelain tea caddy. These were painted with the same motifs as the rest of the tea set and were relatively small, only containing a small amount of tea leaves. There was just enough for a small tea party – the quantity of tea was calculated for six cups. Firstly developed in Meissen, these small porcelain receptacles were in common use during the Rococo period. Their highly rectangular shape with accentuated shoulders was sealed with a small round lid, which was the size of a teaspoon and also served for measuring the correct amount of tea for the pot. Towards the end of the 18th century, the new design style of classicism also revealed itself in the design of the tea set. Now tea caddies were cylindrically round or oval. The accentuated shoulders, small openings and lids stayed but – in contrast to before – were no longer equipped with dainty flowers as a knob but with classical acanthus blossom or pine cones. In the 19th century, the tea caddy reached independence and became an individual piece. It was no longer necessarily part of a particular set and also appeared in individual designs. The receptacles for tea leaves developed as "Tea-Caddy" into a more and more popular gift in England. They were mostly made of silver and in a polygonal shape or as a casket.

Linke Seite
Vasenkollektion „Colletto" von Mercantile, 1982.

Rechte Seite
Aschenbecher für die „Philipp Morris-Künstleredition 2000", jährlich wechselnde Dekore, 1990–2000, Hersteller Rosenthal.

Left-hand side
Vase collection "Colletto" by Mercantile, 1982.

Right-hand side
Ashtrays for the "Philip Morris Artists-Edition 2000", yearly changing decors, 1990–2000, manufacturer Rosenthal.

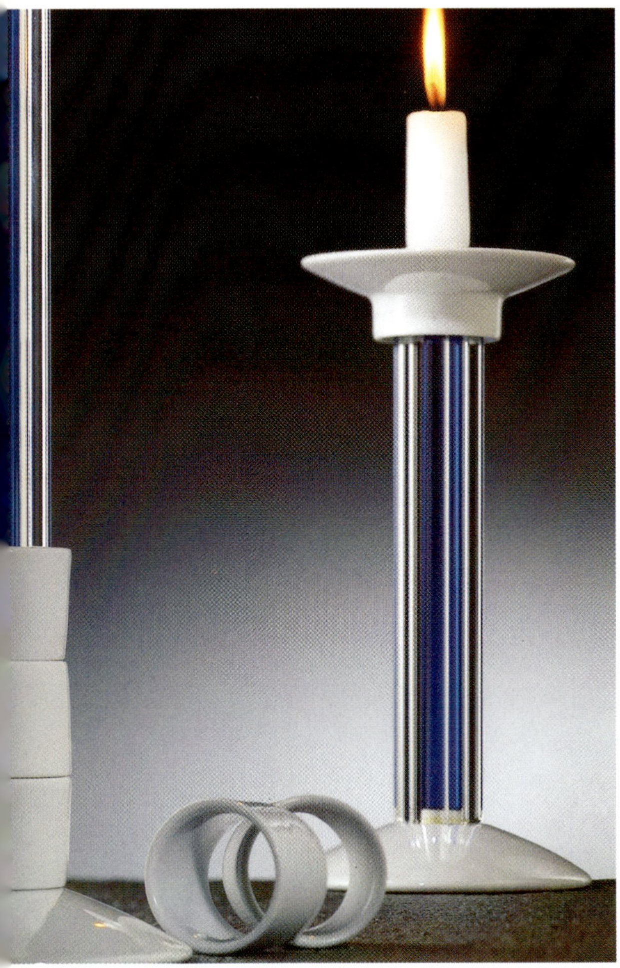

VORWÄRTS IN DIE VERGANGENHEIT

Keines dieser bekannten Vorbilder wählte Tassilo von Grolman für seinen Auftrag von Twinings, sondern er ließ sich vom Ur-Typus ostasiatischer Teebehältnisse aus Porzellan inspirieren. Schon die Größe seiner Teedose lässt alle europäischen Vorbilder vergessen, mit 23,5 cm Höhe ist sie ungewöhnlich groß. Anders als ihre chinesischen Vorläufer gab der Designer seiner Dose keine zylindrische, sondern eine ovoide Form, die an zwei gegenüberliegenden Seiten abgeflacht ist. Dies ist nicht nur ein ästhetisches, sondern auch ein funktionales Gestaltungselement, denn trotz ihrer Größe lässt sich die Teedose dort, wo die Flächen zusammenstoßen, mit einer Hand umfassen und hochheben. Währenddessen kann die andere Hand den ovalen Deckel fassen, der sich zwischen Finger und Handfläche schmiegt. Tassilo von Grolmans Porzellandose ist ein Objekt, das die Sinne anspricht und das man gern zum täglichen Gebrauch in der Nähe weiß. Das Behältnis hat zwei Gesichter, je nachdem, ob man es von seinen Breitseiten oder vom seitlichen Bug her betrachtet. Seitlich wirkt es hoch und schlank, ein Eindruck, der durch den Bug am Korpus noch verstärkt wird. Die breiten Schauseiten vermitteln die Ansicht eines Ovoids, das sich von unten nach oben verbreitert. Die leicht geschwungenen Schultern werden von einem großen ovalen Deckel gekrönt, der die Wölbung nach oben abschließt. Der abgenommene Deckel bildet eine ovale Schale, die von Grolman als ideales Messgefäß für den Inhalt einer Teekanne berechnet hat. Tassilo von Grolmans Teedose ist damit eine kongeniale Schöpfung im Dienst des Teegenusses und des Teezeremoniells zu seinen bereits gestalteten Teegeräten, so zum Beispiel die Kanne „Mono Tea 1" von 1983, die in ihrer Eleganz und innovativen Gestaltung bereits heute zu den Klassikern modernen Designs gehört. Auftraggeber Twinings lässt die Teedose insgesamt zehnmal als Jahresgeschenk in limitierter Auflage von je einem namhaften Künstler dekorieren. So wird bis zum Jubiläumsjahr 2006 eine Kollektion von zehn individuell dekorierten und jeweils einer anderen Teesorte gewidmeten Dosen das von Tassilo von Grolman geschaffene Objekt ins rechte Licht setzen.

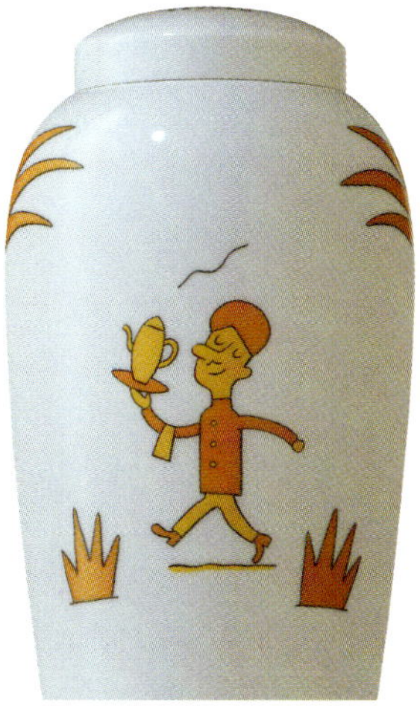

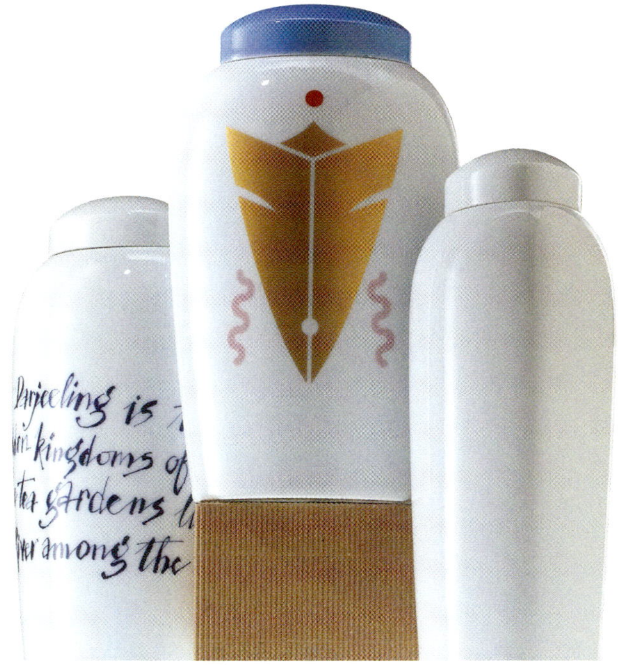

Linke Seite
Kerzenständer mit Serviettenringen für Rosenthal, 1995.

Rechte Seite
Tea-Caddy mit jährlich wechselnden Künstler-Dekoren in limitierter Auflage, 1995.

Left-hand side
Candle holder with napkin rings for Rosenthal, 1995.

Right-hand side
Tea caddy with yearly changing artists' decors in limited edition, 1995.

FORWARDS TO THE PAST

Tassilo von Grolman chose none of these well-known models for his task with Twinings, but rather he was inspired by the prototype of porcelain tea receptacles from Eastern Asia. The size of his tea caddy alone eliminates any reminder of European models – at 23.5 cm high, it is unusually large. In contrast to its Chinese predecessors, the designer did not give his caddy a cylindrical but an ovoid shape, which is flattened on two opposite sides. This is not only an aesthetic but also a functional element of the design – in spite of its size the tea caddy can be held and picked up with one hand at the point where these areas meet. Whilst doing this, the other hand can take hold of the oval lid, which nestles between finger and palm. Tassilo von Grolman's caddy is an object that appeals to the senses and that is a pleasure to keep handy to use every day. The receptacle has two faces, depending on whether it is seen from its broadside or sideways from the bows. From the side it seems tall and slim, an impression that is emphasized further by the bows on the body. The broad sides proffer the view of an ovoid, which widens from bottom to top. The slightly curved shoulders are crowned by a large oval lid, which completes the curve at the top. When removed, the lid forms an oval dish, which von Grolman has intended as the ideal measuring vessel for the content of a teapot. Tassilo von Grolman's tea caddy is thereby a congenial creation in the interests of the enjoyment of tea and the tea ceremony together with his previously created tea utensils – for example the teapot "Mono Tea 1" of 1993 – which already belong to the classics of modern design with their elegance and innovation today. The client, Twinings, is having the tea caddy decorated ten times altogether as an annual gift in limited edition, each time by a notable artist. In this way by the anniversary year of 2006, a collection of ten individually decorated caddies – each dedicated to a different sort of tea – will show the object created by Tassilo von Grolman in its true light.

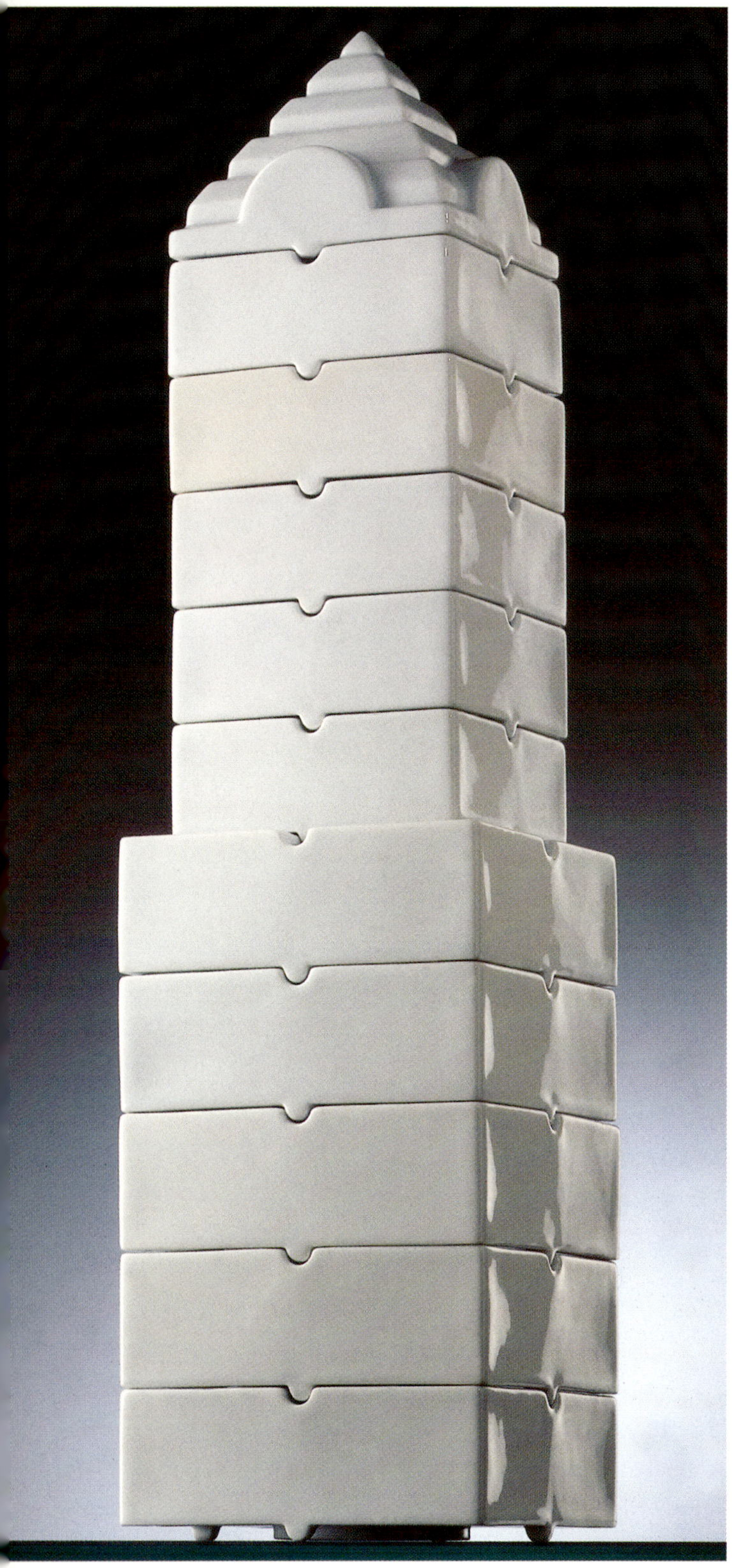

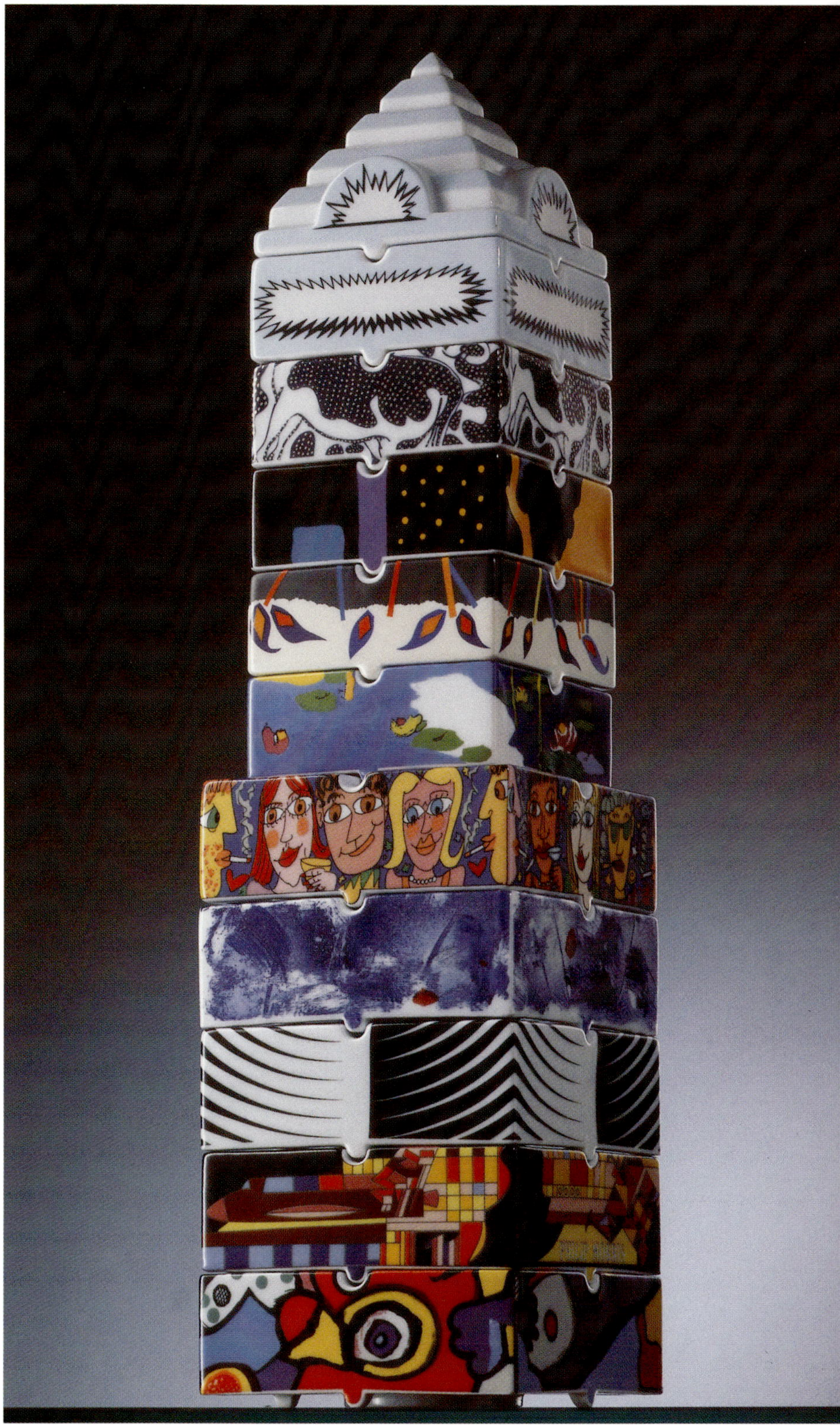

„Philip Morris Künstler-Edition 2000“.
Aschenbecher: Entwurf Tassilo von Grolman,
Dekore von verschiedenen Künstlern.

"Philip Morris Artists-Edition 2000".
Ashtray: concept by Tassilo von Grolman,
decors by various artists.

Frank Stella, 2000

Yang, 1999

Brigitte Doege, 1998

Susan Fateh, 1997

Tadashi Asoma, 1996

James Rizzi, 1995

Attersee, 1994

Marcello Morandini, 1993

Christian Wichmann, 1992

Otmar Alt, 1991

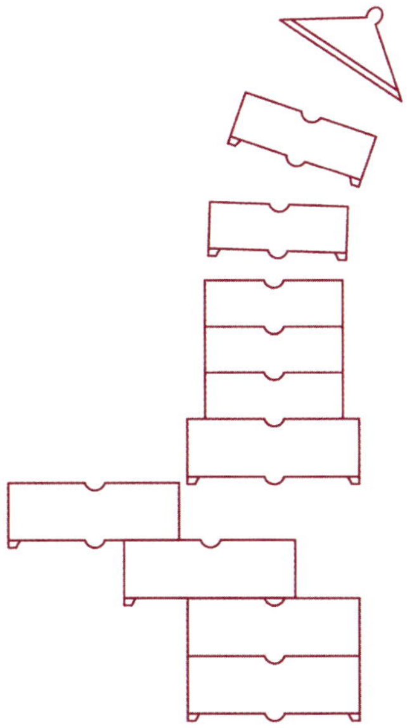

VARIABLES KUNSTWERK

Das zweite von Tassilo von Grolman gestaltete Porzellanobjekt ist ein Kunstwerk, das 1991 begonnen und nach ebenfalls zehn Jahren, im Jahr 2000, vollendet wurde. Auftraggeber Philip Morris wünschte sich wie Twinings zehn Jahresgeschenke für die Freunde des Hauses. Die Präsente sollten sich zur Jahrhundertwende zu einem Gesamtkunstwerk zusammenfügen lassen. So erfand Tassilo von Grolman einen Turm aus zehn stapelbaren Aschenbechern, die in zwei Größen hergestellt wurden. In den ersten fünf Jahren konnten die größeren Ascher gesammelt werden. Sie bilden das Fundament des Turmes. Die kleineren fünf folgten bis zur Spitze. In jedem Jahr dekorierte ein anderer Künstler die dafür vom Designer großflächig gestaltete Wandung der Aschenbecher, wobei jede Seite eine andere Variation des jeweiligen Motivs zeigt. Der Sammler kann sich so seinen Turm aus Aschenbechern in unzähligen Abwandlungen immer wieder neu zusammensetzen, und jedes Mal entsteht ein neues Kunstwerk. Ideenreichtum bei der Gestaltung alltäglicher Dinge ist nur einer der vielen Vorzüge, die die Arbeiten dieses Designers auszeichnen. Welches Material er auch gerade bevorzugt: Das Ergebnis ist immer hervorragendes Design. Asche und Tee – man wünscht sich, dass Tassilo von Grolman noch oft den Werkstoff Porzellan für die Gestaltung dieser und weiterer Themen auswählen wird.

VARIABLE WORK OF ART

The second porcelain object designed by Tassilo von Grolman is a work of art that was begun in 1991 and was also completed after ten years, in the year 2000. Like Twinings, client Philip Morris wanted ten annual gifts for friends of the company. The presents should fit together to form an entire work of art for the turn of the century. So Tassilo von Grolman invented a tower made of ten stackable ashtrays, which were made in two sizes. The larger ashtrays could be collected in the first five years. They form the base of the tower. The five smaller ones followed up to the top. Each year, a different artist decorated the walls of the ashtray, which the designer gave a large surface, whereby every side displayed a different variation of the respective motif. In this way, the collector can reassemble his ashtray tower in countless new variations over and over again, and each time a new work of art comes into being. Richness of ideas in the design of day-to-day things is only one of the many merits, which distinguish the work of this designer. Whatever material he is favouring at the time: the result is always excellent design. Ash and tea – it is to be hoped that Tassilo von Grolman will continue to choose the material porcelain for design of these and other themes.

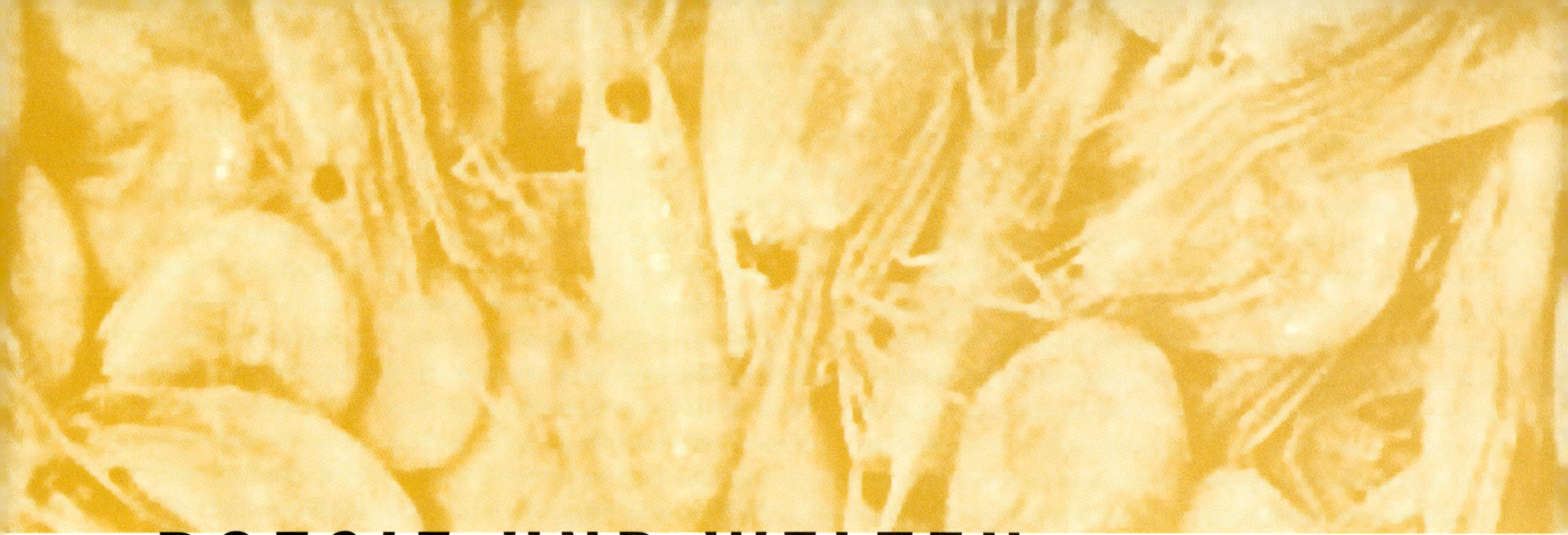

POESIE UND WELTEN

SPAGHETTI MIT KRABBEN

1. Die geschälten King Prawns, Prawns oder kleine Krabben, (je nach Belieben) auftauen und auf Küchenkrepp legen, damit sie relativ trocken werden.

2. Einen Topf mit ca. 5–6 l Wasser aufsetzen, etwas salzen, die Spaghetti in das kochende Wasser geben und ca. 6–10 Minuten kochen (je nach Sorte, steht auf der Packung).

3. Knoblauch schälen, fein hacken und mit Olivenöl und Salz mischen, bis eine klebrige Masse entsteht.

4. Die Frühlingszwiebeln waschen und in ca. 2 cm lange Stücke schneiden. Den Koriander und die Petersilie waschen, schleudern und grob schneiden. Die Pepperoni entkernen, in Längsstreifen und dann in kleine Rechtecke schneiden.

5. In einem Wok oder Topf das Olivenöl erhitzen, die Frühlingszwiebeln, den Knoblauch und die Krabben dazugeben und ca. 3–4 Minuten scharf anbraten. Mit Wein und Fischfond löschen. Die Petersilie, den Koriander und die Pepperoni untermischen und alles 3 Minuten köcheln lassen. Mit Salz abschmecken.

6. Die Krabben mit der Soße über die fertigen Spaghetti auf dem Teller anrichten.

500 g Spaghetti

1000 g tiefgefrorene Krabben

1 Bund Frühlingszwiebeln

1 Bund frischer Koriander

1 Bund Petersilie

3–4 Knoblauchzehen

3–4 rote Pepperoni

0,3 l Fischfond

0,2 l Weißwein

0,1 l Olivenöl

etwas Salz

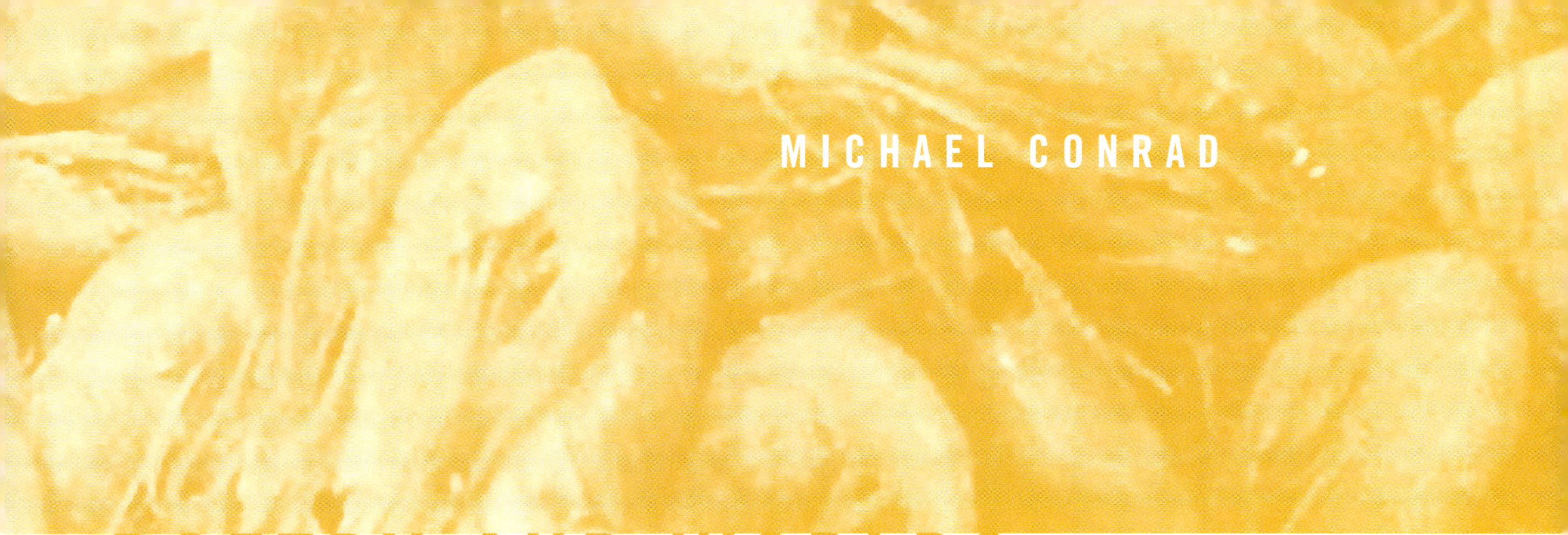

MICHAEL CONRAD

POETRY AND WORLDS

SPAGHETTI WITH SHRIMPS

1 *Thaw the peeled king prawns, prawns or shrimps (according to preference) and place them on paper towels so that they dry somewhat.*

2 *Bring 5–6 l of salted water to a boil, place the spaghetti into the boiling water and cook for 6–10 minutes (exact cooking time depends on the type, see package for details).*

3 *Peel the garlic, chop it finely and mix it with olive oil and salt into a sticky paste.*

4 *Wash the spring onions and cut them into pieces of about 2 cm. Wash and dry the coriander and the parsley and chop them coarsely. Remove the seeds from the peppers, cut them into strips and then into small rectangular pieces.*

5 *In a wok or saucepan, heat olive oil, add the spring onions, garlic and shrimps and braze them for 3–4 minutes. Add the wine and fish stock. Add the parsley, coriander and peppers and simmer for 3 minutes. Add salt to taste.*

6 *Arrange the spaghetti on plates and top them with the shrimps and sauce.*

500 g spaghetti

1000 g frozen shrimps

1 bunch spring onions

1 bunch fresh coriander

1 bunch parsley

3–4 cloves of garlic

3–4 red bell peppers

0,3 l fish stock

0,2 l white wine

0,1 l olive oil

Salt

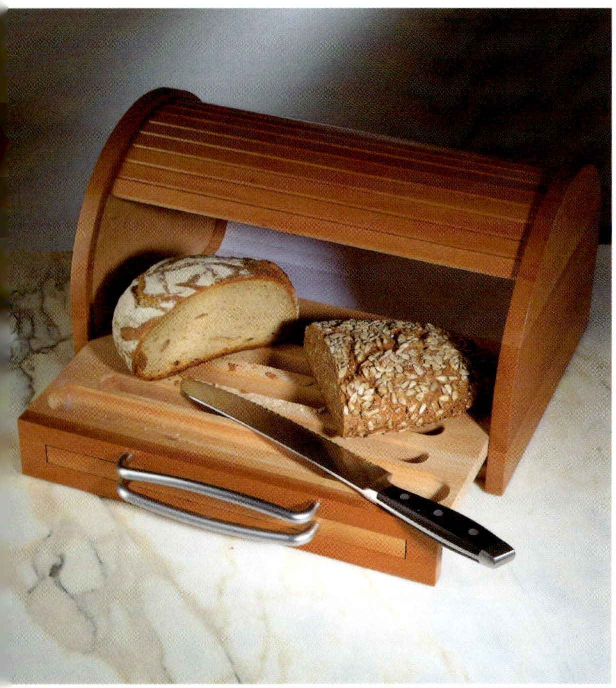

Brotkasten „Krümelmonster" mit integrierter Krümelschublade, Schneidefläche und Messerablage. Mitheis Austria, 1993.

Bread box "Krümelmonster" (crumb monster) with integrated crumb drawer, cutting board and knife receptacle. Mitheis Austria, 1993.

Josepha, unsere Tochter, ist zur Zeit bei uns in Zürich zu Besuch. Sie ist Poetin, also ein Mensch, der Menschen auf die Finger schaut. Sie kennt meinen Freund Tassilo seit 27 Jahren. Ich bat sie, für diesen Beitrag mit mir über Tassilo zu sprechen. Sie willigte ein, unterhält sich aber gerne schriftlich. Und so schoben wir uns das Power Book gegenseitig zu und schrieben Gedanken in den Tassilo hinein.

M Josepha, Tassilos Arbeiten empfinde ich als zeitlos menschlich. Er ist Designer mit ausgeprägter Menschlichkeit. Was sagst Du?

J Wie würdest Du „menschlich" definieren?

M Seine Produkte haben Wärme, Poesie, wollen benutzt werden, man kann sie schmecken, fühlen, hören, sehen, riechen, sie haben Idee, inspirieren zu Gesprächen, sind universell gedacht, sind gut gemacht, zeitlos, addieren einen Wert zum Leben der Leute ..., sind eben besonders menschlich.

J Vor allem sind sie praktisch. Tee ohne Tassilo ist kein Tee. Ich habe immer den Tassilo im Tee, bin aufgewachsen mit den Glasbäuchen und den breiten Sieben, wo sich der Tee entfaltet. Da entfaltet sich auch mancher Gedanke. Tee trinken ist meine Lieblingsbeschäftigung. Wie ist es mit Dir und Tee?

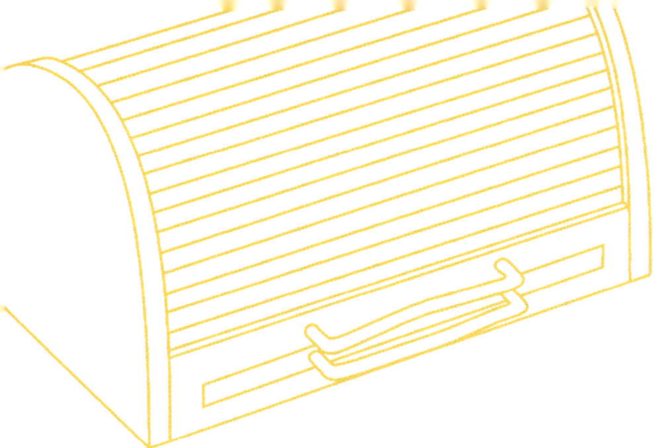

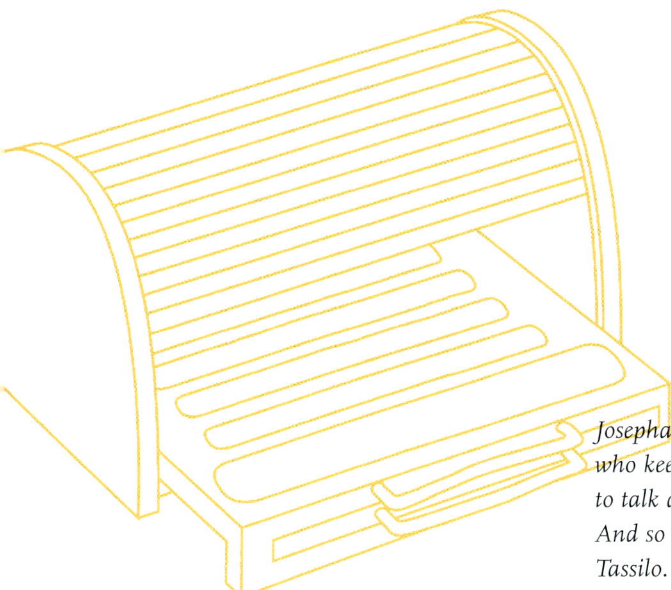

Josepha, our daughter, is visiting us in Zurich at the moment. She is a poet, so a person who keeps a close eye on people. She has known my friend Tassilo for 27 years. I asked her to talk about Tassilo with me for this article. She agreed, but likes to converse in writing. And so we pushed the Power Book to and fro between us and began to write thoughts into Tassilo.

M *Josepha, I feel that Tassilo's work is timelessly human. He is a designer with marked humanity. What do you think?*

J *How would you define "human"?*

M *His products have warmth, poetry, want to be used, they can be tasted, felt, heard, seen, smelt, they have ideas, inspire conversations, are universally intended, well made, timeless, add a value to people's lives..., are just particularly human.*

J *Above all they are practical. Tea without Tassilo is not tea. I always have Tassilo in my tea, have grown up with the bulbous glass body and the wide strainers, where the tea unfolds. Many thoughts also unfold then. Drinking tea is my favourite occupation. What about you and tea?*

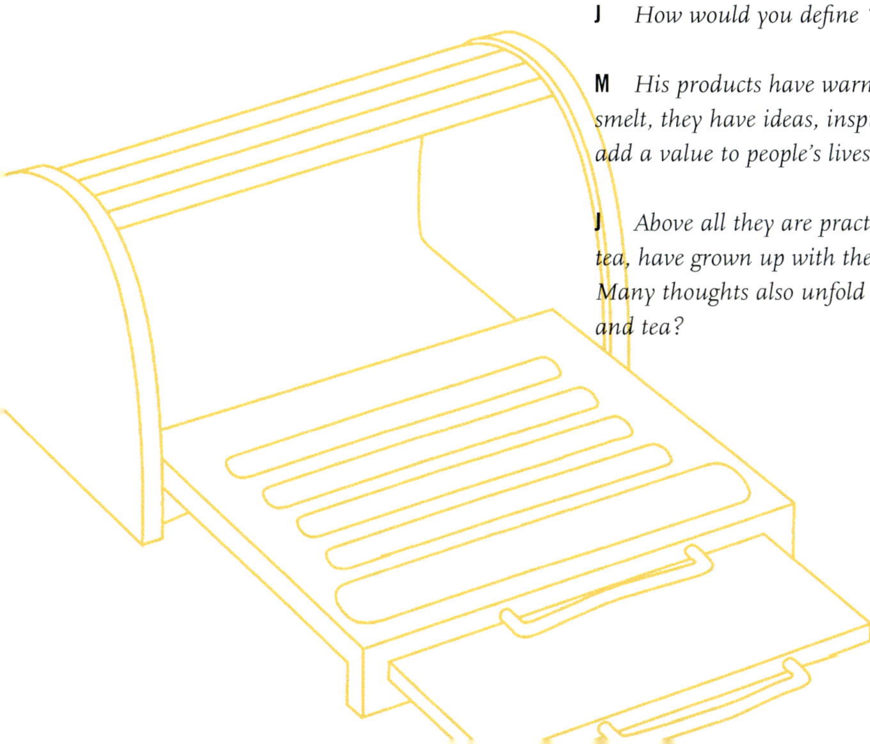

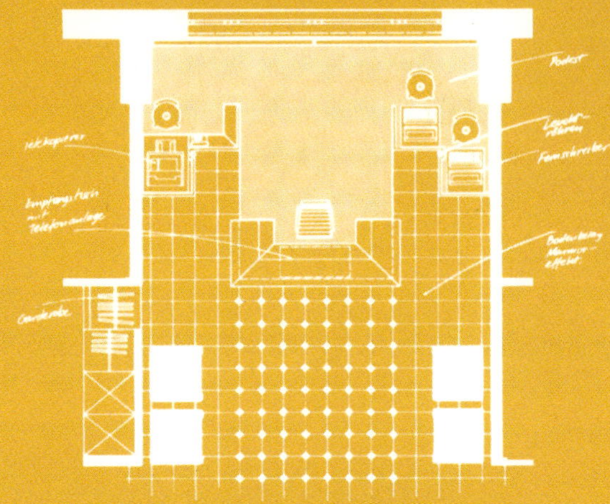
Podest
Fernschreiber
Garderobe

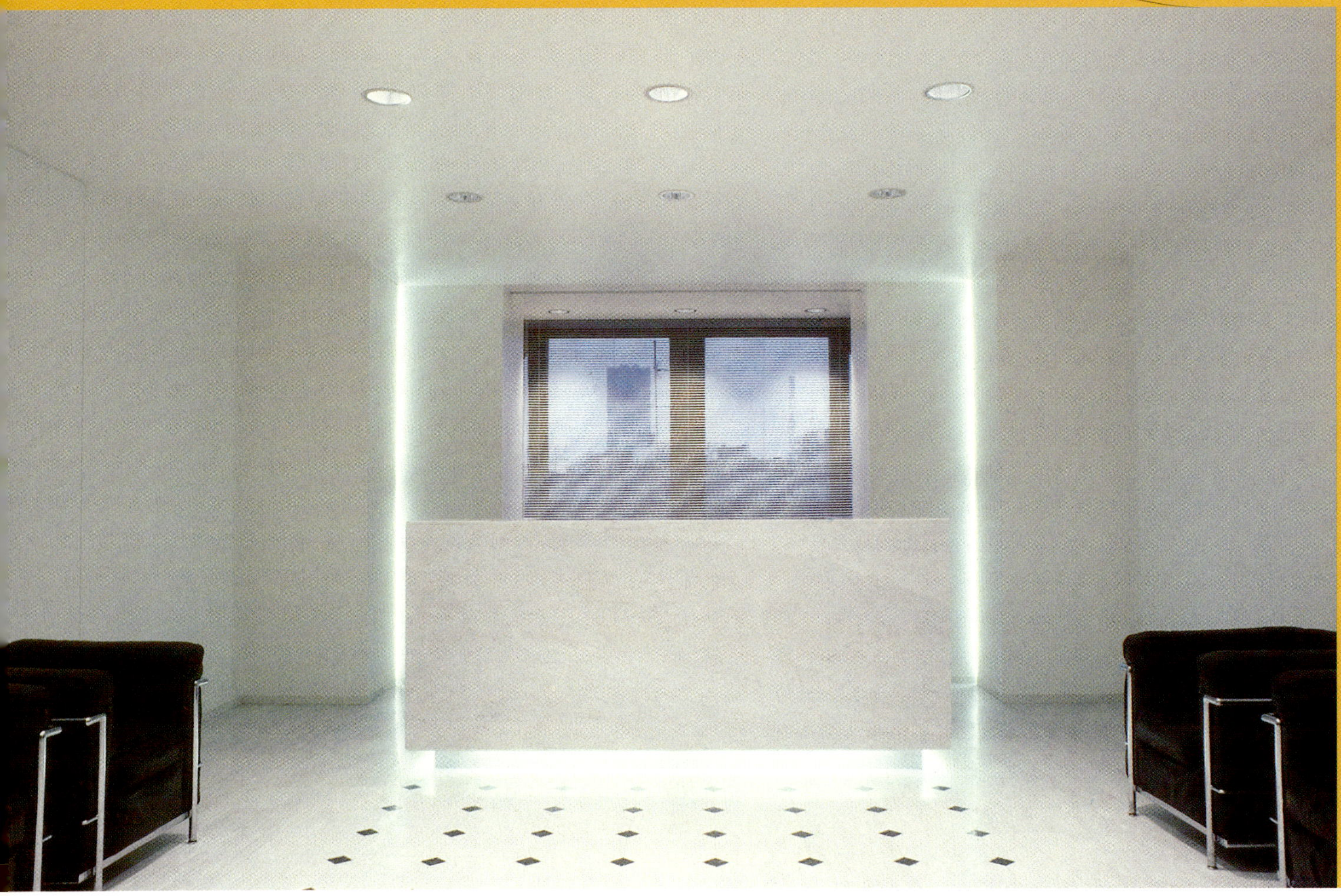

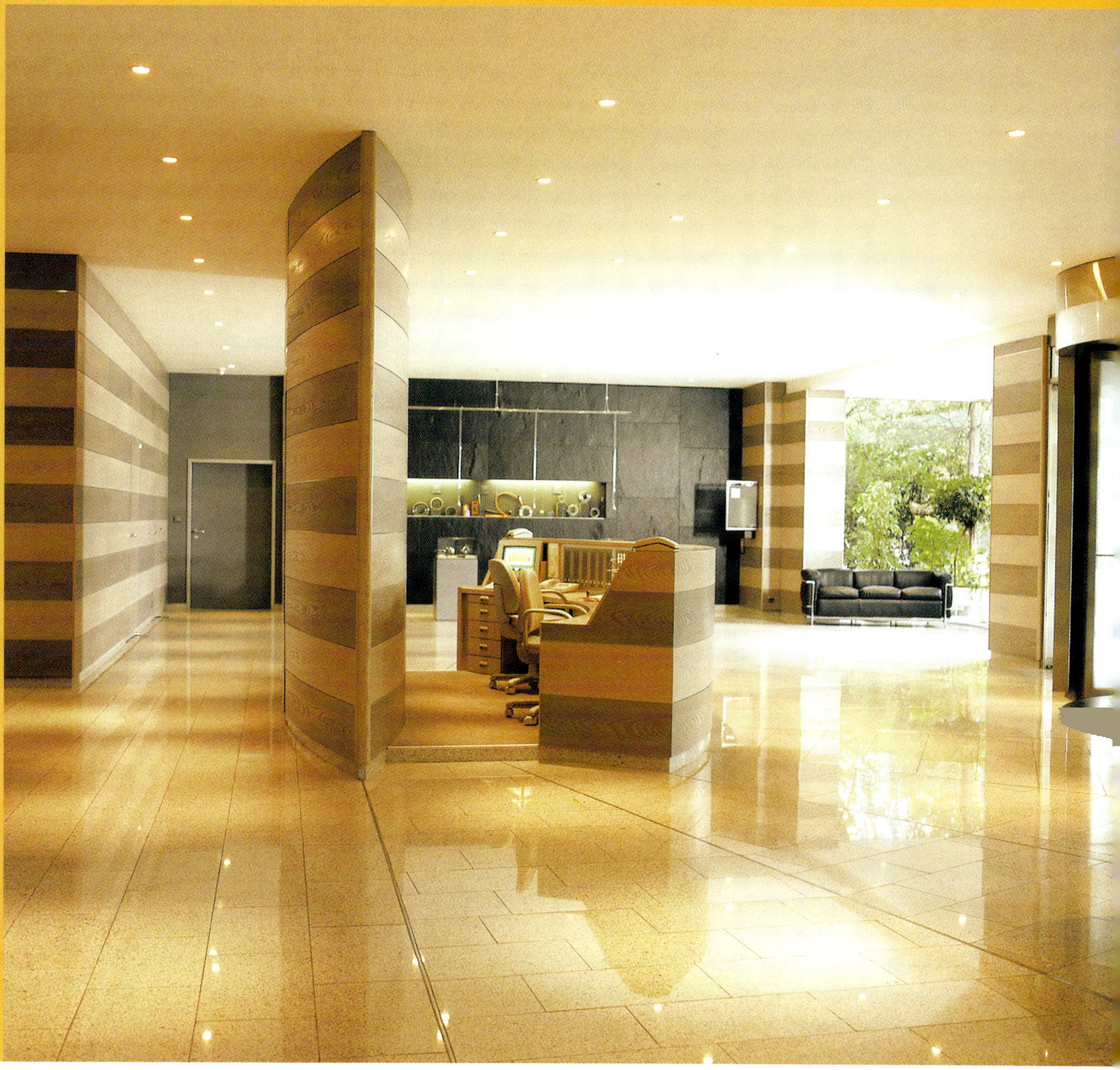

Linke Seite
Gestaltung des Empfangsraumes der Werbeagentur Michael Conrad & Leo Burnett in Frankfurt, 1982.

Rechte Seite
Eingangshalle der SKF GmbH in Schweinfurt, 1989.

Left-hand side
Design of the reception area of the advertising agency Michael Conrad & Leo Burnett in Frankfurt, 1982.

Right-hand side
Entrance hall of SKF GmbH in Schweinfurt, 1989.

M Ich bin kein Teetrinker, ich trinke jede Menge Kaffee. Du hast mit Poesie reflektiert. Poetisch ist auch das Bild, das Deine Schwester Anja von Eurer Oma geschossen hat, das schönste Bild von meiner Mutter. Wäre es, was es ist, ohne Tassilos herzliche Kräutertöpfe?

J Wäre es, was es ist, ohne Oma? Das ist es eben. Man kann Tassilo nicht aus dem Kräutertopf nehmen. Oder Champagner ohne ihn trinken, oder einen Korn schlucken, ohne durch Kimme und Korn zu schauen. Wie war das noch mit dem Bierglas? Ein Produkt mit dem man spielen kann, oder mit dem man Aktivität verbindet, so dass Aktivität und Produkt eins werden, das ist genial.

M Wenn man Produkte kauft, ist es gut, wenn man eine Erklärung für den Kauf hat. Eine Rationalisierung sozusagen. Schnaps kann man aus der Flasche trinken, aus einem Zahnputzbecher, aus einem feinen Wiener-Werkstätten-Glas, aus einem geschnittenen Glas von Lobmeier. Ich war bei Washington Olivetto, einem Werbeguru in Sao Paulo, eingeladen, der mit Tassilos Schnapsgläsern bewaffnet eine grandiose Party garnierte. Man trank dort nicht aus Flaschen, man zelebrierte die Arbeit der Brenner. Wie kommt Tassilo überhaupt in die feinen Kreise von Sao Paulo? Nun, er ist Aristokrat und Seemann.

M *I am no tea drinker; I drink large quantities of coffee. You have reflected with poetry. The picture that your sister Anja took of your grandma – the most beautiful picture of my mother – is also poetic. Would it be what it is without Tassilo's warm-hearted flowerpots?*

J *Would it be what it is without Grandma? That is just it. You cannot take Tassilo out of the flowerpot. Nor drink champagne without him, or swallow schnapps without looking through the bottle. What was that again with the beer glass? A product that you can play with, or with which you connect activity, so that activity and product become one, that is inspired.*

M *When one buys products, it is good to have an explanation for the purchase. A rationalisation, so to speak. You can drink schnapps out of the bottle, out of a tooth mug, out of a fine Viennese-workshop glass, from Lobmeier cut glass. I had an invitation from Washington Olivetto, an advertising guru in Sao Paulo, who garnished a grandiose party armed with Tassilo's schnapps glasses. You did not drink out of bottles there; it was a celebration of the work of the distiller. And how does Tassilo fit into the high society of Sao Paulo at all? Well, he is an aristocrat and seaman.*

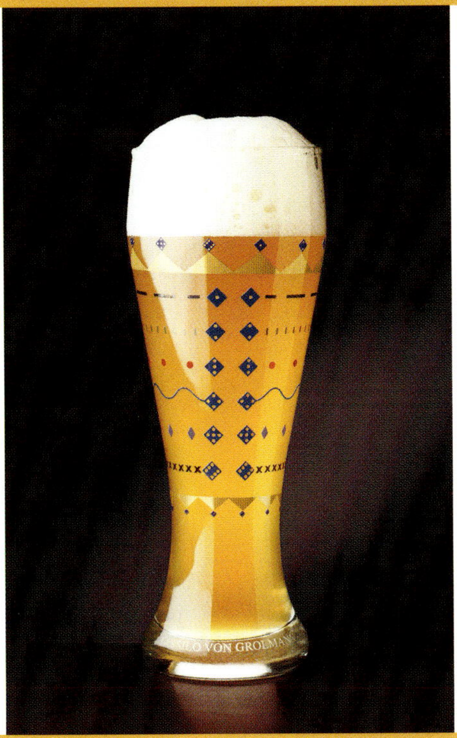

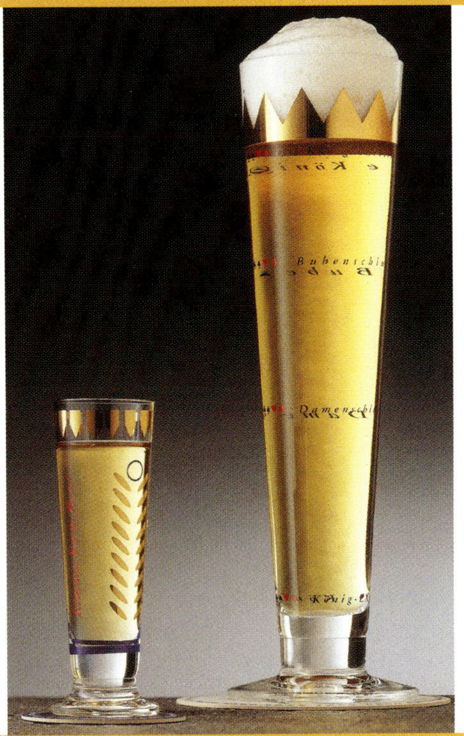

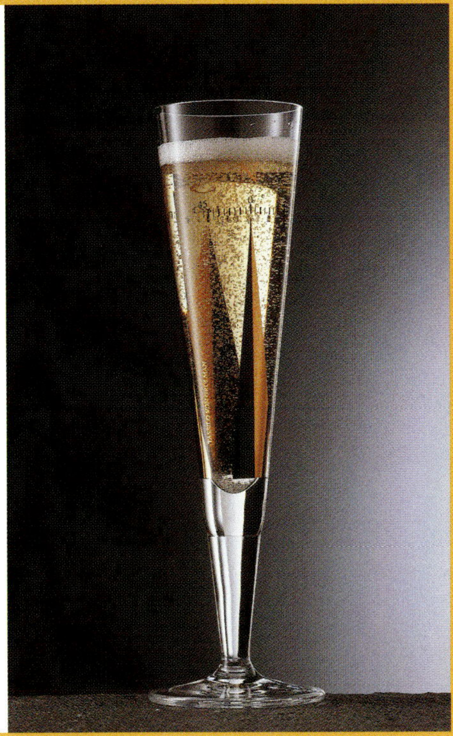

Dekor für Ritzenhoff „Weizenbierglas-Collection", 1997. Dekore für Ritzenhoff „Schnaps- und Bierglas-Collections", 1996. Dekor für Ritzenhoff „Champus-Collection", 1997.

Decor for Ritzenhoff "Wheat-beer Glass-Collection", 1997. Decors for Ritzenhoff "Schnapps and Beer Glass-Collections", 1996. Decor for Ritzenhoff "Champus Collection", 1997.

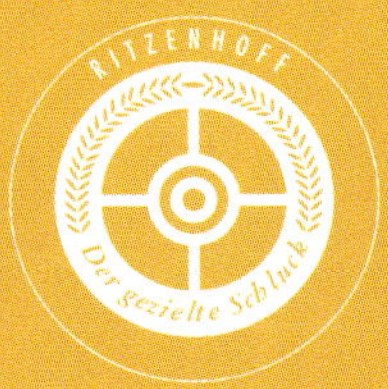

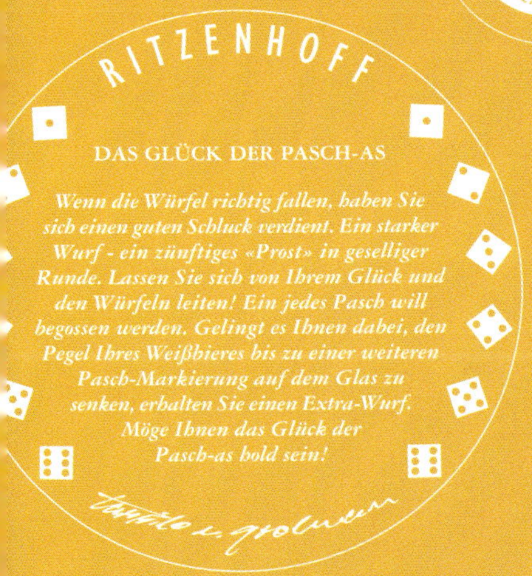

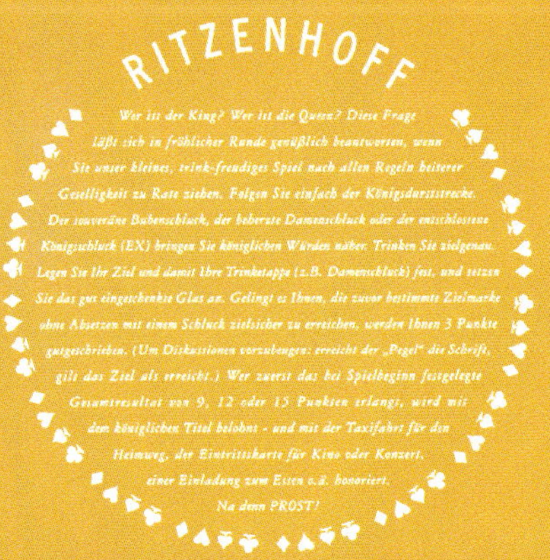

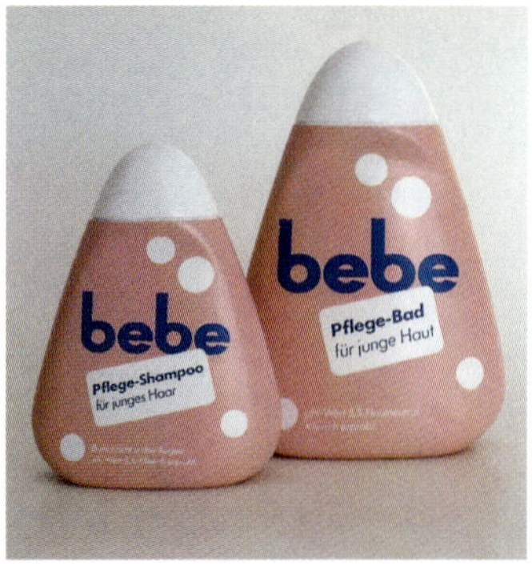

Linke Seite
Verpackungsentwurf für „BeBe" Babykosmetik für Johnson + Johnson, im Team mit Peter Engelhardt, 1987.

Rechte Seite
Einzel- und Doppelwaschtisch für Bad-Möbelprogramme von Allibert. Passgenau durch Kunststein, 1980.

Left-hand side
Packaging design for "Bebe" Babycare for Johnson + Johnson, in cooperation with Peter Engelhardt, 1987.

Right-hand side
Single and twin basins for bathroom furniture by Allibert. Formfitted due to synthetic stone, 1980.

J Aristokrat mit Seidentuch um den Hals. Gespür für das Edle und Kostbare. Eine Kaffee-Thermoskanne, die sich silbern auftürmt und jeden Morgen den Takt wertvoll anschlägt. Seemann, der die Welt erkundet und vor Poseidon und der Unterwelt keine Angst hat. So dringt er ins Unterbewusste und erträumt neue Welten, das Steuerrad fest in der Hand und die Himmelsrichtung im Auge. Seemann-Aristokrat ist die perfekte Vermischung von Gegensätzen, die ein Designer braucht, wenn Gegenstände neu geschöpft werden und gleichzeitig ästhetisch und praktisch sein sollen. So kam Tassilo nach Sao Paulo.

M Ein Mensch mit verwegener Eleganz, ein Abenteurer auf dem Parkett. Als ich Tassilo kennen lernte, trug er feine Lederschuhe, aufgearbeitete englische Sakkos und wohnte in einer Garage, die er als Schlaf- und Wohnzimmer umgebaut hatte. Gehen wir tiefer in den Seemann-Artistokrat.

J So manch ein Seemann findet eine Insel oder eine versunkene Stadt und wird dann König, lässt das blaue Blut fließen, und hundert Jahre später kann sich niemand vorstellen, Aschenbecher zu haben, die sich nicht türmen lassen. Eine Garage kann Insel oder Königreich sein, das hängt von Kreativität und Imaginationskraft ab. Die fehlen dem Tassilo nicht. Erzähle mir doch von der Garage.

M Du machtest das breite Tor auf und warst im Arbeits-, Wohn-, Schlafzimmer, in der Küche und im Klo. Bingo. Schnell drin, schnell draußen. Nichts war weiter als einen Schritt entfernt. Alles was fehlte, war das Auto, das – wie schon erwähnt – durch frischgeputzte Schuhe ersetzt war.

J Durch so einen Raum muss man navigieren können. Da kennt man jede Umdrehung, und trotzdem muss man sich dem Wetter anpassen. Jeder Tag ist anders. Endlose Möglichkeiten eines Tubenverschlusses. Wenn ich mir Töpfe von Tassilo anschaue, oder das Nudelwasser, ohne mich zu verbrennen, abgieße, da weiß ich, dass Tassilo im Topf gewohnt hat, dort noch immer wohnt. Der Flaschengeist.

M Es gibt auch einen akustischen Tassilo. Seine Lachausbrüche sind infektiös. Ich erinnere mich an ein Abendessen in seinem Büro, wo er in einer Ein-Quadratmeter-Küche für 13 Gäste Chinesisches kochte. Über zwei Stunden kam er nicht aus seiner Küche heraus, bat mich schließlich, die weißen Büro-Tische zu decken und zu servieren. Ich wusch die Tische sauber und häufte das Gekochte direkt auf die zwei frisch geputzten Tisch-Platten, ebenso die Saucen. Auch die Kerzen wurden direkt auf die Tischplatte gewachst. Dann riefen wir Tassilo, doch endlich zum Essen zu kommen. Als er realisierte, wie pur sein wunderbares Essen auf dem Tisch um Stäbchen herum garniert war, wurden wir nicht mit einem Schreikrampf bedacht, sondern mit einer gewaltigen Lachsalve, die bis heute sein verbales Markenzeichen ist.

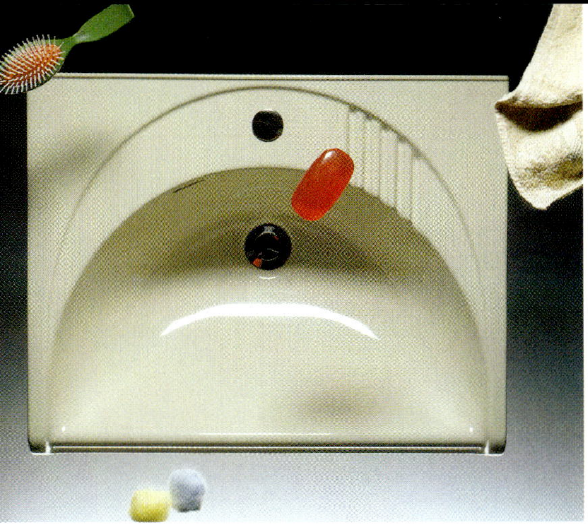

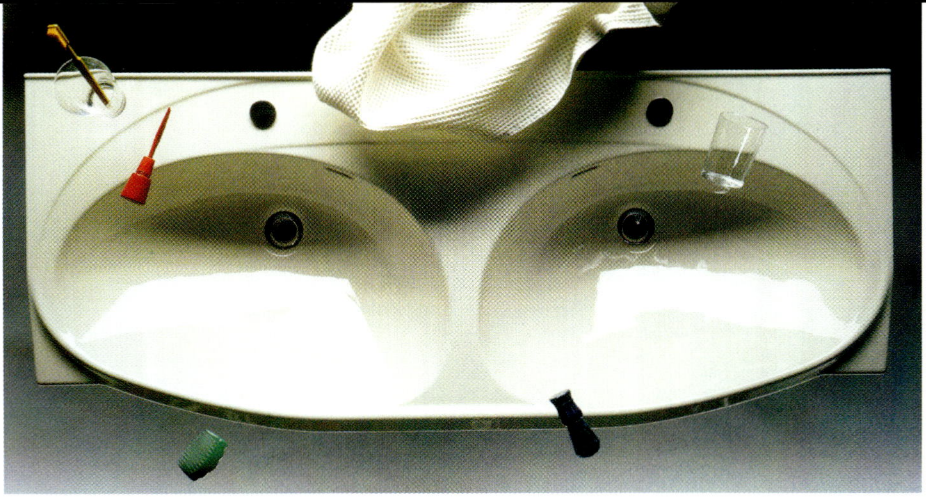

J *Aristocrat with a silk scarf around his neck. A sense for the noble and the precious. A thermal coffee pot, which towers up in silver and every morning valuably strikes up the rhythm. Seaman who explores the world and has no fear of Poseidon or the underworld. Thus he penetrates the subconscious and dreams up new worlds, the ship's wheel firmly in his hand and his eye on the points of the compass. Seaman-aristocrat is the perfect mixture of contrasts, which a designer needs for the creation of new objects that are to be both aesthetic and practical. That's how Tassilo came to Sao Paulo.*

M *A person with bold elegance, an adventurer on the dance floor. When I first got to know Tassilo, he used to wear fine leather shoes, reconditioned English jackets and lived in a garage, which he had converted into a bedroom and living room. Lets go deeper into the seaman-aristocrat.*

J *So many a seaman finds an island or a sunken city and then becomes king, lets the blue blood flow and a hundred years later no one can imagine having ashtrays that cannot be piled up into a tower. A garage can be an island or a kingdom, it depends on creativity and power of imagination. These are not lacking in Tassilo. Tell me about the garage.*

M *You opened the wide door and were in the work, living and bedroom, in the kitchen and in the loo. Bingo. Quickly in, quickly out. Nothing was further than a step away. All that was missing was a car, which – as I mentioned already – was replaced by freshly polished shoes.*

J *You have to be able to navigate through such a room. You know every turn there, and still you have to adapt to the weather. Every day is different. Endless possibilities of a tube cap. When I look at Tassilo's pans, or pour out the pasta water without burning myself, I know then Tassilo has lived in the pot, still lives there. The Genie.*

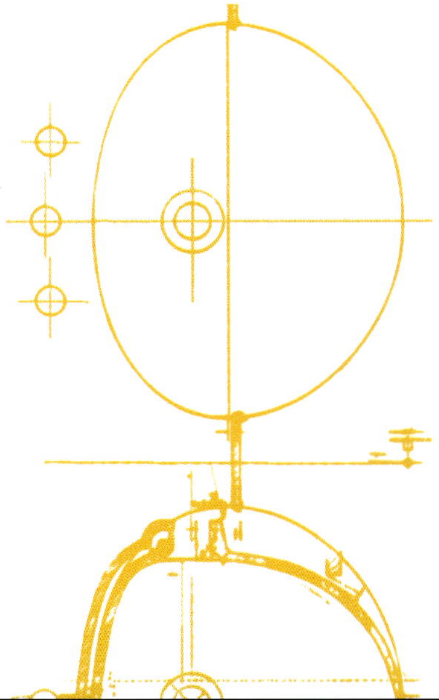

M *There is also an acoustic Tassilo. His outbreaks of laughter are infectious. I remember one dinner in his office, where he cooked a Chinese meal for 13 guests in a one metre square kitchen. He did not come out of the kitchen for over two hours, and asked me finally to lay the white office tables and to serve. I wiped the tables clean and heaped the cooked food directly onto the two freshly cleaned tabletops, as well as the sauces. Even the candles were put straight onto the table top with wax. Then we called to Tassilo to finally come and eat. When he realized how purely his wonderful food was garnished around chopsticks, we were not met with a screaming fit but a powerful volley of laughter, which is still his verbal trademark today.*

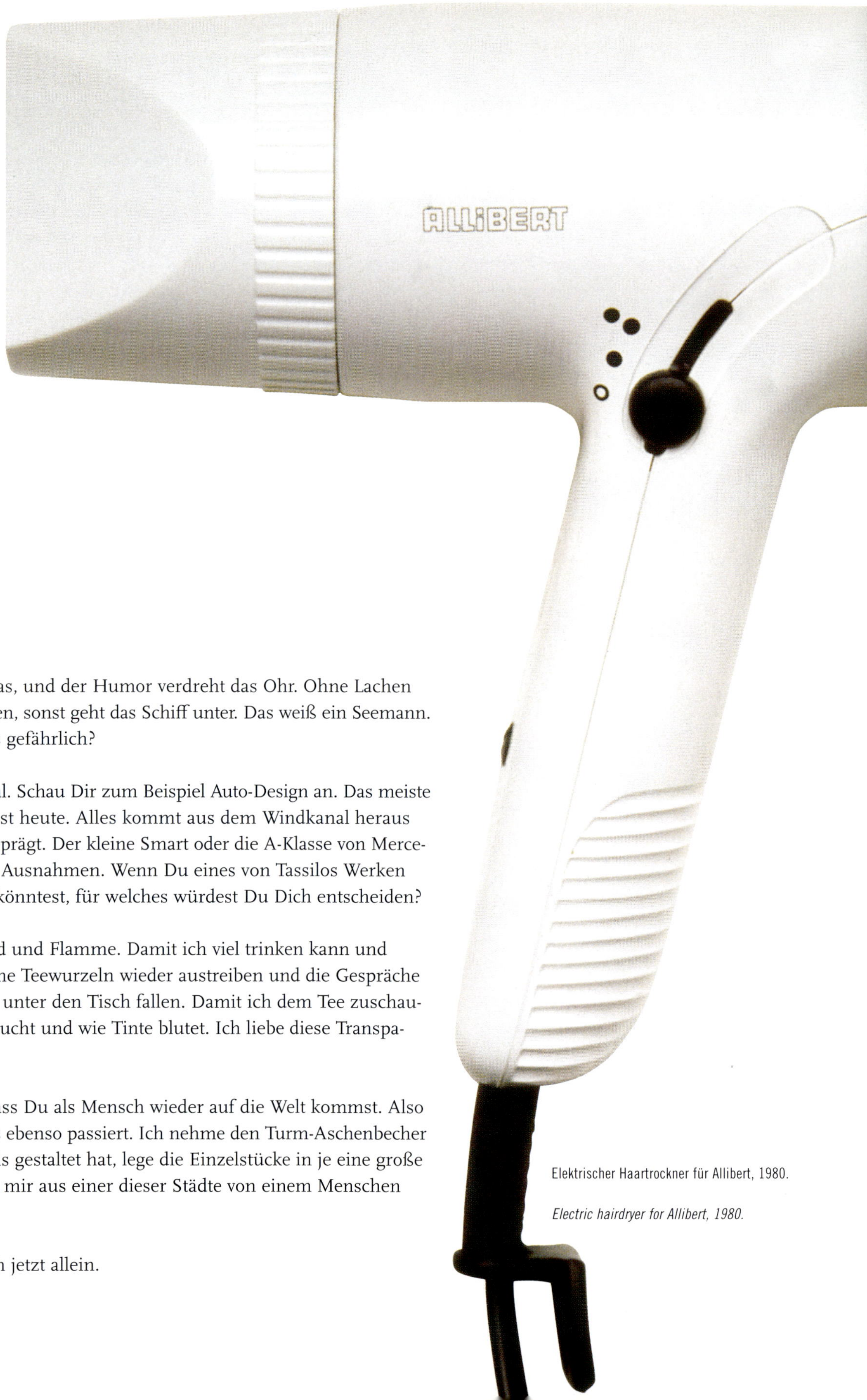

J Lausch mal in ein Tassilo-Glas, und der Humor verdreht das Ohr. Ohne Lachen sollte man keinen Schnaps trinken, sonst geht das Schiff unter. Das weiß ein Seemann. Was das Infektiöse betrifft, ist es gefährlich?

M Für Design ist es fundamental. Schau Dir zum Beispiel Auto-Design an. Das meiste davon ist so infektiös wie die Pest heute. Alles kommt aus dem Windkanal heraus oder ist von Meinungsstudien geprägt. Der kleine Smart oder die A-Klasse von Mercedes sind zwei der ganz wenigen Ausnahmen. Wenn Du eines von Tassilos Werken mit ins nächste Leben nehmen könntest, für welches würdest Du Dich entscheiden?

J Seine Teekanne, mit Rechaud und Flamme. Damit ich viel trinken kann und immer warm bleibe. Damit meine Teewurzeln wieder austreiben und die Gespräche nicht gleich bei der ersten Tasse unter den Tisch fallen. Damit ich dem Tee zuschauen kann, wie er in das Wasser taucht und wie Tinte blutet. Ich liebe diese Transparenz. Und Du?

M Du bist ja ziemlich sicher, dass Du als Mensch wieder auf die Welt kommst. Also gehe ich davon aus, dass mir das ebenso passiert. Ich nehme den Turm-Aschenbecher mit, den Tassilo für Philip Morris gestaltet hat, lege die Einzelstücke in je eine große Hafenstadt und hoffe, dass man mir aus einer dieser Städte von einem Menschen mit Riesen-Lacher berichtet.

J Design, Design, ich lass Dich jetzt allein.

Elektrischer Haartrockner für Allibert, 1980.

Electric hairdryer for Allibert, 1980.

J *Listen into a Tassilo glass and you can hear the humour. You should not drink schnapps without laughter, otherwise the ship will sink. A seaman knows that. But what about the infectious, is it dangerous?*

M *It is fundamental for design. Look at car design for example. Most of it today is as infectious as the plague. Everything comes out of the wind tunnel or is shaped by opinion polls. The little Smart or the A-Class from Mercedes are two of the very few exceptions. If you could take one of Tassilo's creations with you into the next life, which one would you decide on?*

J *His teapot, with warmer and candle. So that I can drink a lot and always keep warm. So that my tea roots sprout again and the conversations do not immediately fall under the table with the first cup. So that I can watch the tea, how it dives into the water and bleeds like ink. I love this transparency. And you?*

M *You certainly are quite sure that you are coming back into the world as a human. So I assume that the same will happen to me. I am taking the tower-ashtrays with me, which Tassilo designed for Philip Morris: I will put each individual piece in a sea port, and hope that I hear reports from these towns of a person with an enormous laugh.*

J *Design, design, I now leave you alone.*

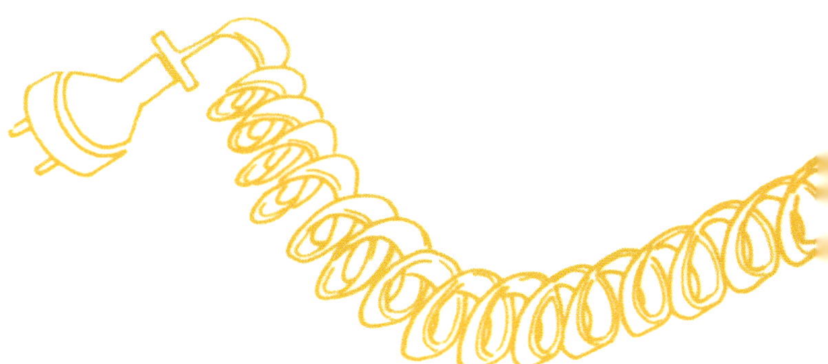

NACHGEDANKE

Der abgeschnittene Rock von Mary Quandt.
Der designte Sound von George Martin.
Gefilmte Musik von MTV.
Obi-Wan-Kenobi's von George Lucas.
Kunsthaus-Hauskünste von Frank Gehry.
Gesurfte Schriften von David Carson.
Farbige Äpfel von Steven Jobs.
Blaues Pepsi gegen rotes Coke.
Backsteine und Mörtel gegen das World Wide Web.
Die alte Wirtschaft gegen die neue.
Die Welt, in der wir leben,
von Wettbewerb geprägt wie nie zuvor.
Auftritt. Image. Look. Logo. Branding, global.
Design ist schon lange keine grafische Sache mehr.
Design ist Denken,
holistisch integrierter Teil von Geschäftsideen,
Produktideen, Markenideen, Kommunikationsideen.
Die menschlichsten Gedanken, Ideen, Designs
sind immer die erfolgreichsten.
Tassilo von Grolman hat sich gut positioniert.

AFTERTHOUGHTS

The miniskirt of Mary Quant.
The designed sound of George Martin.
Filmed music of MTV.
Obi-Wan-Kenobi of George Lucas.
Art house-household arts of Frank Gehry.
Surfed writings of David Carson.
Coloured apples of Steven Jobs.
Blue Pepsi against red Coke.
Bricks and mortar against the World Wide Web.
The old economy against the new
The world in which we live,
shaped by competition as never before.
Entrance. Image. Look. Logo. Branding, global.
Design has not been one graphic thing for a long time now.
Design is thinking,
holistically integrated part of business ideas,
product ideas, brand ideas, communications ideas.
The most human thoughts, ideas, designs
are always the most successful.
Tassilo von Grolman has positioned himself well.

Schriftgestaltung „Jagro" und Verpackungsgestaltung für „Apomed Analogpräparate", 1988.

Font design "Jagro" and package design for "Apomed Analog-Preparations", 1988.

JAGRO

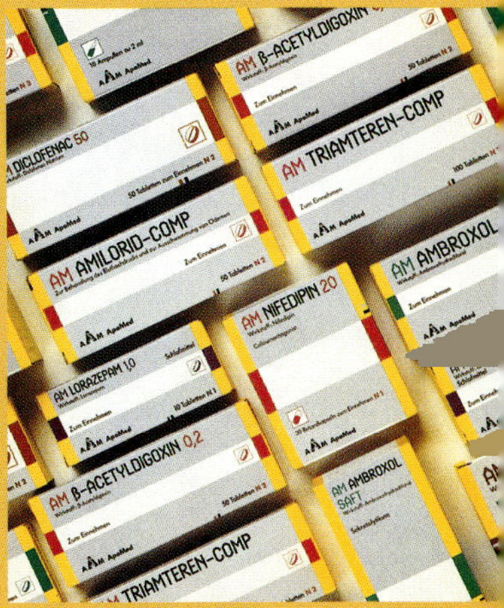

ABCDEFGHIJK
LMNOPQRSTUVWXYZ
ÄÖÜÆŒ
abcdefghijklmnopqrs
tuvwxyz
äöüßæœàâéêèôû
1234567890

MITARBEITER/EXTERNE PARTNER
STAFF/EXTERNAL PARTNERS

Tassilo von Grolman Design besteht aus einem Team aus fünf Mitarbeitern und dem Office Management. Darüber hinaus wird es durch Praktikanten in- und ausländischer Designschulen unterstützt, die ein studienbegleitendes Semester an konkreten Projekten absolvieren.

Ein Team externer Produkt-Designer, Entwicklungs- und Fertigungskonstrukteure, Modellbau-Spezialisten für Muster- und Prototypenbau, Grafik-Designer, Illustratoren und Fotografen wird jeweils projektbezogen eingesetzt.

„Professionalität, Engagement und Teamgeist sind die Basis für unseren Erfolg. Und der Erfolg ist immer eine Leistung des Teams. Dafür danken wir an dieser Stelle insbesondere den Mitarbeitern, die gestern und heute gemeinsam unsere Projekte zum Erfolg geführt haben." Tassilo von Grolman

Tassilo von Grolman Design consists of a team of five and the office management. In addition, students doing a period of practical training from design schools within Germany and abroad, who are completing one term of study working on a specific project, provide support.

A team of external product designers, development and production designers, model building specialists for the making of samples and prototypes, graphic designers, illustrators and photographers is brought in as each project dictates.

"Professionalism, commitment and team spirit are the basis of our success. And our success is always a team achievement. At this juncture, we are particularly grateful to past and present colleagues who have together led our projects to success." Tassilo von Grolman

Hubertus von Allwörden
Christian Altmann
Richard Bachinger
Horst Baerenz
Thomas Bech
Alexander Beck
Wolfgang Bloch
Jürgen Bode
Anja Conrad
Anke Delingat
Willi Demel
Peter Engelhardt
Christoph Felger
Matthias Felsch
Catarina Ferreira
Heinrich Fiedeler
Siggi Fischer
Jörg Gambke
Rudi Geiger
Kent Gerber
Olivera Gligoric
Vera Götz
Joop Greypink
Elke Griech
Bernhard Gruhlke
Jörg Hänel
Mihoko Hanzawa
Martin Hauenstein
Florian Heine
Sonja Hermann
Julia Heukelbach
Clemens Hilger
Oliver Hill
Suat Ilkiz
Steffen Imhof
Alexandra Jaeckel
Helmut Jahn
Erika Koch
Mirko Krizanovic
Verena Kuhn
Anke Lambrecht
Norbert Latocha
Klaus Leuschel
Frank Lieser
Annabell Loewe
Inge Maisch
Klaus Malorny
Manfred Martini
Gabriele Maßmann
Ulrich Mattner
Melanie Megel
Axel Meise
Stefanie Minner
Bernd Mirbach
Isabel Mittl
David Neuner-Jehle
Peter Oberhaus
Melanie Osswald
Birgit Osterhage
Robert Pfann
Tobias Plath
Tabea Pötsch
Sabine Raab
Christine Renner
Telse Richter
Annette Rust
Werner Ruwald
Janine Schaphorst
Olaf Schilgen
Klaus Schultheiß
Marie-Luise von Seelen
Kichinosuke Shikuwa
Gilbert von Sohlern
Ralf Stede
Dagmar Steffen
Sören Strayle
Lutz Thiele
Michael Thiem
Wolfgang Tscherneff
Wolfgang Usbeck
Nina Virus
Horst Wackerbarth
Stefan Weil
Thomas Wilkens
Harald Windecker

KUNDEN/PROJEKTE
CLIENTS/PROJECTS

Das Designbüro Tassilo von Grolman kann eine beeindruckende Zahl von Klienten vorweisen. Über 120 verschiedene Unternehmen beauftragten in den letzten 25 Jahren die Tassilo von Grolman Design GmbH. Neben den sehr unterschiedlichen Einzelprojekten unterhält von Grolman aber auch langjährige Kundenbeziehungen. Er ist ein Global Player, Auftraggeber kommen nicht nur aus Deutschland, sondern auch aus Dänemark, Frankreich, Großbritannien, Indonesien, Irland, Italien, Japan, den Niederlanden, Österreich, der Schweiz und den USA. Tassilo von Grolman Design verfügt über ein vielfältiges Leistungsangebot, nämlich Produkt-, Packungs- und Grafikdesign, Corporate Design, Messe- und Architekturdesign, Designberatung. Einige der im folgenden aufgeführten Unternehmen wurden im Laufe der Zeit in andere Unternehmen integriert bzw. bestehen nicht mehr unter den hier genannten ursprünglichen Firmennamen.

The Tassilo von Grolman design company can boast an impressive number of clients. Over the past 25 years, over 120 different companies have commissioned Tassilo von Grolman Design GmbH. Alongside the very varied individual projects, von Grolman also has client relationships stretching back many years. He is a global player. Work providers do not only come from Germany but also from Denmark, France, Great Britain, Indonesia, Ireland, Italy, Japan, the Netherlands, Austria, Switzerland and the USA. Tassilo von Grolman Design offers a wide range of services including product design, packaging design, graphic design, corporate design, exhibition and architectural design as well as design consultancy. Over the course of time, some of the companies listed below have been integrated into other companies or no longer exist under the original names mentioned here.

Alfi GmbH
Allibert GmbH
Arzberg
BASF
Berndes KG
BfG Bank AG
Birkel Sonnen Bassermann Nahrungsmittel GmbH
Bodum
Bredemeijer
Carl Mertens GmbH
Clifford Chance Pünder
Club English Tea GmbH
Commerzbank/Hausinvest
Compas Mediaagentur
Corning
De Beers
Deutsche Bahn AG
Deutsche Bank AG
Deutsche Forschungsgemeinschaft
Deutscher Designer Club
DEVK Versicherungen
Diebold Deutschland GmbH
Dr. Scheller Durodont GmbH
Eckes
Eduscho
Ellen Betrix Cosmetic
Emform GmbH
Erco
Erdal Rex
Fachverband Baukeramik
General Foods
Gillette
Glaskunst Lauscha GmbH
Halle Münsterland
Henkel
Hewi
Hoechst AG
IDE International Design & Entertainment GmbH
Ideal Standard
Jacobs-Suchard-Gruppe
Jade Cosmetic GmbH
Junita Fruchtsaft oHG
Kaltenberg Brauerei
Kopp GmbH
L'Oréal
Lidl & Schwarz Warenhandel GmbH
Löwenbräu AG
Lufthansa Flight Training GmbH
Manhattan Cosmetics GmbH
Mattel GmbH
M & C Schiffer GmbH
Mercantile GmbH
Merck Manan GmbH
Michael Conrad & Leo Burnett Werbeagentur
Mitheis oHG
Moët-Hennessy Deutschland GmbH
Mono-Metallwarenfabrik
Montblanc
Nassauische Sparkasse
Nestlé Deutschland AG, Maggi GmbH
Orang Tua Personal Care Division
Philip Morris GmbH
Plaudo Italia
Rhône-Poulenc
Ritzenhoff Cristal AG
Rösle-Metallwarenfabrik GmbH
Rosenthal Porzellanfabrik AG
Salomon, S.A.
Schenk Fahrzeugwerk GmbH
Schlumberger Dehm & Zinkeisen
Scholl-Werke GmbH
Schott-Zwiesel
Seagram Deutschland GmbH
SKF GmbH
Spectra-Physics
Spring AG Metallwarenfabrik
Tipp-Ex GmbH
Tobu Department Store Company Ltd.
Topdeq GmbH
Troika Böll & Cie
Vereinigte Papierwerke
Verkehrsamt der Stadt Frankfurt am Main
Vileda GmbH
Villeroy & Boch
Warsteiner Brauerei
Wega Radio GmbH
Wella
Woolworth Deutschland GmbH

PROJEKTE
PROJECTS
Stand: Oktober 2000
Status: October 2000

PRODUKTDESIGN
PRODUCT DESIGN
Bad-Ausstattungsserie „Lavela“
Bad-Sanitär-Objekte/Design-Konzept
Besteck-Serie/Design-Konzept
Besteck aus Kunststoff/Design-Konzept
Bewegungsmelder „Mouse Watch“
Bügel „Gürtelschwan“
Brotkasten „Krümelmonster“
Büro-Accessoires Emform „Wave“
Champagnergläser „Christiane“/„Iphigenie“
für Künstler-Edition Kulturstadt Weimar
Champagner-Kühler „Moët & Chandon“
Dauerwand-Kalender/eigene Entwicklung
Designers Bear Collection Ritzenhoff „Drachen-Bär“
Designers Bear Collection Ritzenhoff „Snoozy Bär“
Duschen und/oder Baden auf kleinem Raum/Design-Wettbewerb
Einhand-Mischbatterie/Design-Konzept
Einweg-Geschirr/Design-Konzept
Fensterwischer „Vileda 2 in 1“
Fusselrolle/Design-Konzept
Gaszählerregler „Dehm & Zinkeisen“
Geschirr „Espresso Grande“/Design-Konzept
Gewürze aufbewahren/Design-Konzept
Gewürzschrank/Design-Konzept
Glas aus Eis „Glacilo“, eigene Entwicklung
Glaskanne und Teegläser „Boral“
Haartrockner „Allidry Studio“
Haartrockner & Schnellheizer „Allibert-Thermodry“
Hand-Daten-Terminals
Heizungs- und Klimaanlage-Elemente für PKW
Isolierkanne Alfi „Achat“
Isolierkanne Alfi „Aura“
Isolierkanne Alfi „Big Mama“
Isolierkanne Alfi „Change“
Isolierkanne Alfi „HotelDesign“
Isolierkanne Alfi „Juwel-Tee“
Isolierkanne Alfi „Lipstick“
Isolierkanne Alfi „Night & Day“
Jahrespräsent-Aschenbecher-Serie „Philip Morris Edition 2000“
Kaffeebereiter Mono „Cafino“ 0,75 l
Kaffeebereiter Mono „Cafino-Color“ 0,75 l
Kaffeebereiter Mono „Cafinero“ 0,75 l
Karosserie-Design für ein Automodell mit optoelektronischer Ausstattung
Kaviar-Servierschale Mono „Filio“
Kehrmaschine/Design-Konzept
Kerzenleuchter-Serie „Oestergaard“
Kerzenleuchter-Serie „Rosenthal“ mit aufsteckbaren Serviettenringen
Korrekturstift „Tipp-Ex“/Design-Konzept
Küchenmesser-Serie/Design-Konzept
Küchenrollenhalter
Lampen-Kollektion, eigene Entwicklung
Marmeladengläser Mono „Filio“
Mikrowellen-Geschirr/Design-Konzept
Modulares Standbau-System als Bar-Equipment/Design-Konzept
Multi-Kühl-Bar „Alfi“
Porzellan-Serie „Arzberg“/Design-Konzept
Präsentationsdisplay für Marlboro Design Shop
Präsentationsregal für SB-Handel
Prellbock-Anlage in Kopfbahnhöfen/Design-Konzept
Récamiere, Design-Objekt für das internationale Projekt Kunst, „Design und Barbie“
Saft-Dispenser „General Foods“
Schalter-Programm „Kopp“
Schaufensterdisplay für Werbung „Volksbanken-Raiffeisenbanken“
Schließfach-Anlage für Bahnhöfe
Schoko-Fondue-Set „Mono“
Schreibtisch-Accessoires „Montblanc“/Design-Konzept
Shop-in-Shop-System für Trademarketing Promotion
Spardose Deutsche Bank „Prisma“
Spritzsieb/Design-Konzept
Tea Caddy „Twinings Collection 2006“
Teekanne Bodum „Chambord“
Teekanne Mono „Tea 1“
mit passendem Rechaud, 1,5 l
Teekanne Mono „LAB 2“
mit integriertem Rechaud, 1,5 l
Teekanne Mono „CAT“, 1,5 l
Teekanne Mono „Filio“
mit passendem Rechaud, 1,5/0,6 l
Teekanne Mono „Filio“
mit integriertem Rechaud, 1,5 l
Teekanne Mono „Tee-Color“, 1,5 l
Teekanne Mono „Santé“, 1,5/0,6 l
Teetassen Mono „Filio“ mit Zucker- und Sahneset
Thekendisplay für Deutsche Bank
Top-Gastronomie-Ausstattungsserie „Philip Morris Light American“

Topfserie Berndes „Bonanza“
Topfserie Spring „Compact GLI“
TV-Gerät „Wega“
Urbanes Design, Ideen-Wettbewerb
Vasen-Kollektionen „Mercantile“
Waschbecken für Badmöbel-Programme
Elektrischer Wasserkessel/eigene Entwicklung
Zahnbürsten „Durodont“
Zahnbürsten „Formula“
Zigarrenaschenbecher „Hommage à Marianne Brandt“
Zigarrenaschenbecher „Sancho“

GRAFIKDESIGN
GRAPHIC DESIGN
A 340-Flugsimulator von Lufthansa
Broschüre für Beauty Studio Ellen Betrix Cosmetic
Broschüren für Cerha, Hempel & Spiegelfeld
Broschüren zum Mumm-Design-Workshop
Broschüre für Jubiläum NHT & Partner Architektengemeinschaft
Broschüren für „Philip Morris Edition 2000“
Broschüren für Pünder, Volhard, Weber & Axster
Broschüren für Tea Caddy „Twinings Collection 2006“
Büchergestaltung zum Wettbewerb „Designer bewerten Design“
Dekor Milchglas-Collection von Ritzenhoff
Dekor Schnapsglas-Collection von Ritzenhoff
Dekor Bierglas-Collection von Ritzenhoff
Dekor Weizenbierglas-Collection von Ritzenhoff
Dekor Teeglas-Collection von Ritzenhoff
Dekor Champagnerglas-Collection von Ritzenhoff
Dekor Aqua-Glas-Collection von Ritzenhoff
Dekor Herbs-Kräutertopf-Collection von Ritzenhoff
Giveaway „Frankfurter Grüne Soße“, Poster des Verkehrsamts der Stadt Frankfurt
Grafischer Auftritt Woolworth Deutschland
Marken-Logos Calsiwol 800/Steinwolle 1000
Medaille und Urkunden für Deutsche Forschungsgemeinschaft
Logo und Geschäftspapiere der DEVK Versicherungen/Design-Konzept
Logo und Geschäftspapiere für Dolzer Masskonfektionäre/Design-Konzept
Logo und Geschäftspapiere für Grolman & Grolman Marketing- und Projektmanagement GmbH
Logo und Geschäftspapiere für Moritz Hunzinger Public Relations
Logo und Geschäftspapiere Stoffel & Partner Anwaltssozietät in Zürich
Pop-up Weihnachtskarte „Tannenbaum“ für UNICEF
Produkt-Prospekte für Mono Metallwarenfabrik
Salesfolder für „The feature series of Design Life“ von Philip Morris
Salesfolder für Marlboro Film Division
Salesfolder für V-Dia Fotofachlabor
Salesfolder „Wall Street-Journal“ für die Wall Verkehrswerbung
Wort-Bildmarke für Viviq Demenz-Präparat/Design-Konzept

PACKUNGSDESIGN
PACKAGING DESIGN
Alkoholfreies Bier, Etiketten/Design-Konzept
Amaryl Diabetes-Präparat/Design-Konzept
Apomed Analog-Präparate
Bad Pyrmonter Mineralwasser, Etiketten
BASF-Chrome Super II-Cassette-Gehäuse/Design-Konzept
Belsoft Badezusatz, Flaschenkörper
Benedict-Fertiggerichte Produkt-Sortiment
Birkel's No. 1 Nudel-Range
Biovital, Flaschenkörper SB-Flasche
Chandon Extra Seco, Geschenkverpackung
Chandon Brut, Geschenkverpackung
Criss-Tropical Drink/Design-Konzept
Dermomild Waschlotion, Flaschenkörper
Diètina-Fruchtnektar Flaschen- und Etiketten-Design
Dr. Kousa-Weizenkleie/Design-Konzept
Durodont-Zahnbürsten
Durodont Kinder-Zahnbürsten
Duschdas-Range/Design-Konzept
Eckes-Nektar, Etiketten/Design-Konzept
Ellen Betrix Augenkosmetik
Goebel Glas-Produkte, Geschenkverpackungen
Hennessy XO, Geschenkverpackung

Hennessy VS, Geschenkverpackung
Hotel im Wasserturm Köln, Hotel-Accessoires
Hotel Imperial Annecy, Hotel-Accessoires
Jade Weizenkeimöl-Kosmetik-Range, Etiketten
Junita-Fruchtsaft, Flasche und Etiketten
Lavela Bad-Ausstattungsserie
Lensex-Haushaltsreiniger, Flaschenkörper mit Sprühmechanismus
Lidl & Schwarz, Verpackungen für Eigenmarken
L'Oréal-Haarschaum-Serie, Etiketten/Design-Konzept
L'Oréal-Schaumfestiger-Serie, Etiketten/Design-Konzept
Maaloxan Magen-Tonikum, Flaschenkörper und Etiketten/Design-Konzept
Maggi-Tiefkühlkost/Snacks/Design-Konzepte
Manhattan Cosmetics Gesamtsortiment
Milton Desinfektionsmittel, Flaschenkörper
Mono-Metallwarenfabrik, Verpackungen für Teekannen
Noril-Badreiniger mit Sprühmechanismus, Re-Design Flaschenkörper
Öko-Bier, Flaschenkörper/Design-Konzept
Papermate, Klipp-Geschenkboxen
Philip Morris Zigaretten-Verpackungen/Design-Konzepte
Robino-Bad-Accessoires
Schneider, Verpackungen für Großbild-Objektive
Scholl-Fußpflege-Produkte, Re-Design Gesamtsortiment
Scholl-Flexibelle-Gesundheitssandale, SB-Verpackung
Scholl-Fitness-Sandale, SB-Verpackung
Silkience-Haarpflegeserie, Etiketten
Silverstream irisches Mineralwasser, Flaschenkörper und Etiketten/Design-Konzept
Süßstoff-Behälter/Design-Konzept
Tempo-Taschentücher, verschließbare Box
Terracotta-Kräutergarten „Grüne-Soße"
Twinings Collection 2006, Geschenkverpackung
Twinings Iced Tea, Geschenkverpackung
Wega-Radio- und TV-Geräte, Produktsortiment
Wella „Neopon" Haarshampoo, Flaschenkörper
Wella „Recrin" Haarshampoo, Flaschenkörper
Wella „Reginal" Haarshampoo, Flaschenkörper
Woolworth-Eigenmarke „Caressa"-Haushaltshelfer, Logo- und Packungsgestaltung
Woolworth-Eigenmarke „Solano"-Glühbirnen, Logo- und Packungsgestaltung
Woolworth-Eigenmarke „Tornac"-Batterien, Logo- und Packungsgestaltung

CORPORATE DESIGN
CORPORATE DESIGN

Cerha, Hempel, Spiegelfeld, Anwaltssozietät
Clifford Chance Pünder, internationale Anwaltssozietät
Jacobi, Metzger, Furkel, Anwaltssozietät
Nassauische Sparkasse/Design-Konzept
Pünder, Volhard, Weber & Axster, internationale Anwaltssozietät
PVW-Wirtschaftsprüfungsges.mbH
Rhône-Poulenc Apomed-Analog-Präparate
Schenk Fahrzeugwerk GmbH
Thieleking & Marko, Anwaltssozietät

DESIGNBERATUNG
DESIGN CONSULTANCY

„BfG goes Retail", Bank für Gemeinwirtschaft
„Maggi-Edition", Europäischer Design-Wettbewerb, Nestlé Deutschland Maggi GmbH
„Mumm-Design-Workshop 2000", Seagram Deutschland
Aktionskonzepte für Chivas Regal, Seagram Deutschland
„Design Today", Tobu-Department Store Tokio

ORIENTIERUNGS-LEITSYSTEME
GUIDING SYSTEMS FOR ORIENTATION

für ein Krankenhaus in Zusammenarbeit mit Erco in Lüdenscheid
für die Halle Münsterland in Münster
für das Firmengelände der SKF GmbH in Schweinfurt

ARCHITEKTUR-DESIGN
ARCHITECTURAL DESIGN

Ausstellungen „Preisbewußtes Bauen" für Deutsche Bank-Zweigstellen
Ausstellungen „Kleincomputer auch in Ihrem Betrieb" für Deutsche-Bank-Zweigstellen
Ausstellungen und Preisverleihungen „Marlboro Design Förderpreis" in deutschen Großstädten
Ausstellungen des Mumm-Design-Workshop, Konzeption, Koordination und Realisation
Dekorations-System für Dresdner Bank/Design-Konzept
Empfangs- und Konferenzbereich für Admenting Werbeagentur
Eingangs-, Empfangs- und Konferenzbereich für Werbeagentur Bode, Glebocki, Thelen
Eingangs-, Empfangs- und Konferenzbereich für Compas Mediaagentur
Neugestaltung Gebäude, Empfangs- und Konferenzbereich für Contur Werbeagentur
Eingangs-, Empfangs- und Konferenzbereich für Diebold Deutschland GmbH
Empfangs- und Konferenzbereich für Lürzer, Conrad Werbeagentur
Eingangs- und Empfangsbereich für Werbeagentur McCann-Erickson Werbeagentur
Empfangsbereich für Michael Conrad & Leo Burnett Werbeagentur
Eingangs- und Empfangsbereich für SKF GmbH
Empfangsbereich für J.W. Thompson Werbeagentur
Konferenz-Bereich für Schott-Glaswerke/Design-Konzept
Event-Räume für Preisverleihung Philip-Morris-Forschungspreis
Event-Räume für Werbegipfel München
Laserscape-Projekt für Frankfurt/Design-Konzept
Messestand für Birkel Sonnen-Bassermann/Design-Konzept
Messestand für Datapoint Software
Messestand für De Beers/Design-Konzept
Messestand Fachverband Baukeramik
Messestand Junita Fruchtsaft oHG
Messestand König Ludwig Brauerei Kaltenberg
Messestand für Philip-Morris-Forschungspreis
Messestand für Pro Sieben
Messestand für Red Box Verlag
Messestände für Salomon Sportartikel
Messestand für Schott Zwiesel
Mobiles Messe-Standbau-System für SKF GmbH
Messestand für Spectra-Physics Laser
Messestand für Wall Verkehrswerbung
Neuer Auftritt Kaufhaus Woolworth Zeil/Frankfurt
Showroom für Glas-Präsentation Schott-Zwiesel
Showroom für Topdeq GmbH

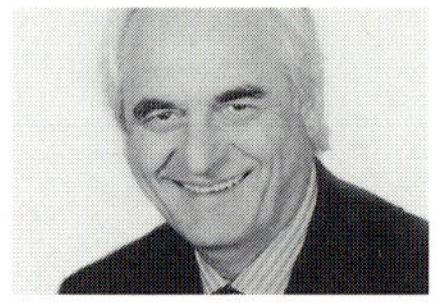

THOMAS FEICHT

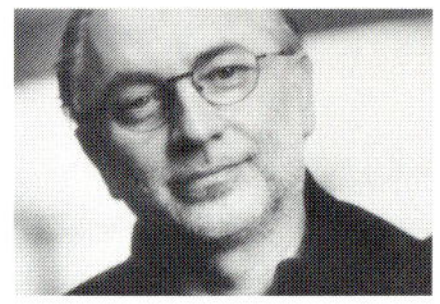

MICHAEL CONRAD

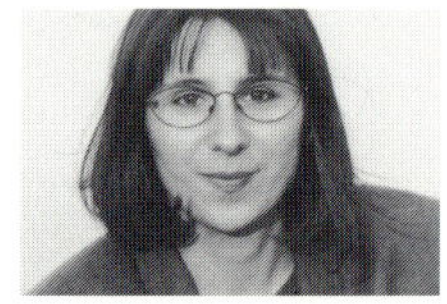

MARION GODAU

AUTOREN
AUTHORS

MICHAEL CONRAD
ist einer der erfolgreichsten deutschen Kreativen der internationalen Werbeszene und weltweit begehrtes Jurymitglied. 1968 startete er in Frankfurt/Main bei Young & Rubicam, war Geschäftsführer bei Heumann Ogily & Mather und TBWA und gründete 1975 gemeinsam mit Walter Lürzer die Werbeagentur Lürzer, Conrad, die 1980 eine Partnerschaft mit Leo Burnett eingingen. Seit 1997 ist Michael Conrad der Chefkreative und Vice Chairman der weltweit operierenden Werbeagentur Leo Burnett Company, Chicago.

is one of the most successful German creatives of the international advertising scene and is sought after worldwide as a member of jury panels. He started out in 1968 in Frankfurt with Young & Rubicam, was managing director at Heumann Ogily & Mather and TBWA and in 1975 founded the advertising agency Lürzer, Conrad, together with Walter Lürzer, which entered into a partnership with Leo Burnett in 1980. Since 1997, Michael Conrad has been Head Creative and Vice-Chairman of the advertising agency Leo Burnett Company, Chicago, which operates worldwide.

THOMAS FEICHT
Nach einer Schriftsetzerlehre studierte er Visuelle Kommunikation in Mainz und arbeitete unter anderem Ende der siebziger Jahre als Art-Director bei Lürzer, Conrad in Frankfurt/Main. 1985 gründete er die Werbeagentur Trust, 1997 Instant Corporate Culture. Thomas Feicht war jahrelang im Vorstand des Art Directors Club von Deutschland und des Deutschen Designer Clubs und ist heute Präsident des DDC.

After an apprenticeship in typesetting, he studied visual communication in Mainz, and amongst other things worked as Art Director with Lürzer, Conrad in Frankfurt. He founded the advertising agency Trust in 1985, and in 1997 Instant Corporate Culture. For years, Thomas Feicht was a committee member of the Art Director Club of Germany and of the Deutscher Designer Club (German Designer Club) and is today President of the DDC.

MARION GODAU
studierte Industrie-Design und Soziologie in Berlin und lehrt(e) Designtheorie und -geschichte, u.a. an der Fachhochschule Hannover und der Hochschule der Künste Berlin. Sie ist freie Designautorin, Publizistin und Designconsultant. Marion Godau lebt in Berlin.

studied industrial design and sociology in Berlin and taught (and teaches) theory and history of design at the Hanover College and College of Arts in Berlin amongst others. She is a freelance author on design, publicist and design consultant. Marion Godau lives in Berlin.

KATJA KRETZSCHMAR
studierte Kunstgeschichte (Schwerpunkt Design) und Film- und Medienwissenschaft in Erlangen und München und arbeitet heute als Fachjournalistin für Zeitungen und Zeitschriften. Sie ist freie Mitarbeiterin der form-Zeitschrift für Gestaltung und form.de.

studied history of art (majoring in design) and film and media studies in Erlangen and Munich, and today works as a trade journalist for newspapers and magazines. She works on a freelance basis for "form", and "form.de".

RENATE PUVOGEL
In Freiburg, Rom, Hamburg, Kiel und Aachen studierte sie Kunstgeschichte, Germanistik und Geographie und war von 1973 bis 1996 wissenschaftliche Mitarbeiterin am Suermondt-Ludwig-Museum in Aachen. Seit 1973 schreibt sie als freie Journalistin in zahlreichen Zeitschriften und Katalogen über zeitgenössische Kunst. Renate Puvogel lebt in Aachen.

She studied history of art, German and geography in Freiburg, Rome, Hamburg, Kiel and Aachen, and was an assistant at the Suermondt-Ludwig-Museum in Aachen from 1973 to 1996. Since 1973 she has written for many newspapers and catalogues on contemporary art. Renate Puvogel lives in Aachen.

KATJA KRETZSCHMAR

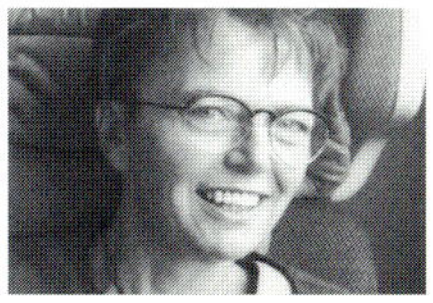

MICHAEL SCHNEIDER

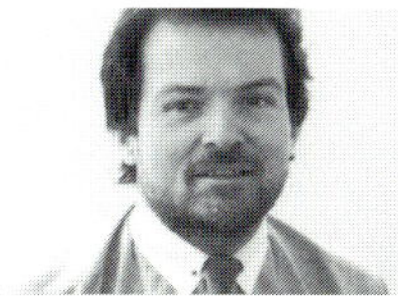

BEATRIX FREIFRAU VON WOLFF METTERNICH

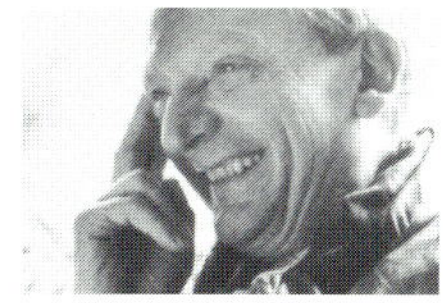

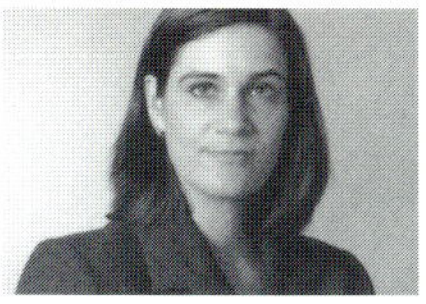

RENATE PUVOGEL

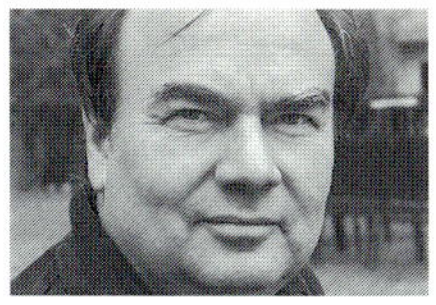

HANS-ULRICH VON MENDE

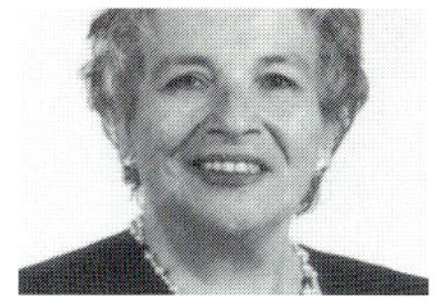

KURT WEIDEMANN

MICHAEL SCHNEIDER
ist seit 1976 Geschäftsführer des „Instituts für Neue Technische Form“ in Darmstadt. Seitdem konzipiert, organisiert, gestaltet und realisiert der Architekt und Bildhauer Ausstellungen mit dem Schwerpunkt Industriedesign, daneben freie Arbeiten und Gestaltungsaufgaben in Architektur und Design sowie Beratung von Firmen und Kommunen.

has been Managing Director of the "Institut für Neue Technische Form" (Institute of new technical design) in Darmstadt since 1976. Since then the architect and sculptor has conceived, organised, designed and carried out exhibitions with a focus on industrial design. He also undertakes freelance work and creative projects in architecture and design as well as consultancy for companies and communities.

HANS-ULRICH VON MENDE
studierte Architektur und ist seit 1990 Partner des Architekturbüros „NHT & Partner“ in Frankfurt/Main. Er schreibt in zahlreichen Zeitschriften, Büchern und Zeitungen über Autodesign, seit 1976 regelmäßig in der „Frankfurter Allgemeinen Zeitung“. Hans-Ulrich von Mende lebt in Frankfurt/Main.

studied architecture and has been a partner in the architectural firm "NHT & Partner" in Frankfurt since 1990. He writes in many magazines, books and newspapers about automotive design, and has been a regular contributor to the "Frankfurter Allgemeine Zeitung" newspaper since 1976. Hans-Ulrich von Mende lives in Frankfurt.

BEATRIX FREIFRAU VON WOLFF METTERNICH
studierte in Wien und Freiburg. Seit 1974 ist sie Kustodin des Museums der Porzellanmanufaktur Fürstenberg, von 1989 bis 1999 war sie dort Direktorin. Die promovierte Historikerin veröffentlichte zahlreiche Publikationen zur Geschichte der Porzellankunst.

studied in Vienna and Freiburg. She has been custodian of the museum of porcelain manufacturing in Fürstenberg since 1974, and she was director there from 1989 to 1999. A doctor of history, she has published numerous written works on the history of porcelain art.

KURT WEIDEMANN
gehört seit vielen Jahren zu den renommiertesten Kommunikationsdesignern in Deutschland. Er ist sowohl bekannter Grafiker und Typograph – unter anderem für DaimlerChrysler und die Deutsche Bahn – als auch Designberater und Publizist. Kurt Weidemann lebt und arbeitet in Stuttgart.

has for many years been one of the most renowned designers in the field of communication in Germany. He is both a well-known graphic designer and typographer – for DaimlerChrysler and the Deutsche Bahn amongst others – and a design consultant and publicist. Kurt Weidemann lives and works in Stuttgart.

BIBLIOGRAPHIE/FOTONACHWEIS
BIBLIOGRAPHY/PHOTO CREDITS

VERÖFFENTLICHUNGEN VON TASSILO VON GROLMAN
PUBLICATIONS OF TASSILO VON GROLMAN

ARTIKEL/ARTICLES

Bachinger, Richard (Hg.): Unternehmenskultur/D 1990. Darin: Unternehmenskultur und Verpackungsdesign
DDC (Hg.): Das deutsche Designbuch No. 1, Vorwort/D 1993
DDC (Hg.): Das deutsche Designbuch No. 2, Vorwort/D 1996
Frankfurter Allgemeine Zeitung/D 11.8.89, 9.2.90, 21.2.90, 16.3.90, 23.3.90, 12.4.90, 23.11.90
Getränkeindustrie/D 5/91
Horizont/D 3/91
Verpackungsrundschau/D 11/90
werben und verkaufen/D 4/89, 6/90, 8/90, 9/90, 7/91, 11/91, 7/92
ZEIT-Punkte/D 5/97

VERÖFFENTLICHUNGEN ÜBER TASSILO VON GROLMAN UND TASSILO VON GROLMAN DESIGN
PUBLICATIONS ON TASSILO VON GROLMAN AND TASSILO VON GROLMAN DESIGN

BÜCHER/BOOKS

Besser sein – wettbewerbsfähig mit Design/ 1995
Design Agencies Europe 1/D 1992
Design im 20. Jahrhundert/D 1989
Design in Hessen/D 1993
Design-Innovationen '83/D 1983
Design-Innovationen '89/D 1989
Design-Lexikon Deutschland/D 2000
Design-Report 3/D 1994
Design und Wohnen/D1993
Deutsche Standards/D 1995
Edizioni Condé Nast, Germania'/I 1993
Ein Kochbuch von Designern/D 1995
Euro Design Guide/D 1992
German Design/D 1990
Graphis Product Design 1/CH 1994
Graphis Packaging Design 7/CH 1996
Il tempo degli oggetti, gli oggetti nel tempo/I 2000
Inspiration, International Design Manual/D 1992
Inspiration, International Design Manual/D 1993
Maggi Edition – Europäischer Designwettbewerb/ D 1995
Messestände, Temporäre Architektur für die Präsentation von Industrieprodukten und Dienstleistungen/D 1987
Modern Design 1890–1990/USA 1990
Neue Standuhren im städtischen Raum/D 1995
Sushi No. 2 ADC-Nachwuchswettbewerb 99/D 1999
Tee, der Guide für Kenner und Geniesser/ Tee, die kleine Schule/D 1997
Top Design des 20. Jahrhunderts/D 1999

ARTIKEL/ARTICLES

AIT/D 5/92
Ambiente/D 2/92
Arcade/D 11/94
Der Tagesspiegel/D 2.4.96
Die Bar/D 3/93
Die Schaulade/D 4/92, 2/93, 12/94
Die Vitrine/D 6/90
essen & trinken/D 2/84
Forme/I 101/83
Frankfurter Allgemeine Zeitung/D 5.9.92
Getränkeindustrie/D 5/91
Graphik Visuelles Marketing/D 06/84
Horizont/D 4/90
Industriemagazin/D 6/89
Metal Times/D 3/90, 6/91
md/D 2/85, 5/93
Neue Verpackung/D 8/90
Neue Zeit/D 18.3.93
Objekt/D 9/92
Porzellan + Glas/D 3/89, 4/93
Report/D 1/77
Süddeutsche Zeitung Magazin/D 18.6.93
Vernissage/D 1/94
werben und verkaufen/D 12/87, 4/88, 1/89, 8/90, 3/92, 11/93
Wirtschaftswoche/D 11/92
Zeitmagazin/D 9/90
Zuhause Wohnen/D 1/92, 4/93

VORTRÄGE/SEMINARE
LECTURES/SEMINARIES

Concurso Jovem Designer. Lissabon/Po 1991
Corporate Identity als Klammer von Design und Werbung. Rat für Formgebung, Frankfurt am Main/D 1988
Der Luxus des Einfachen oder: Designer zwischen Anspruch und Wirklichkeit. Museum für Moderne Kunst, Frankfurt am Main/D 1992
Erfahrungsbericht aus der Arbeit eines Designers. Bauhaus Dessau/D 1991
Highlighting – Lust auf Lack aus der Sicht des Designers. DuPont Deutschland GmbH, Düsseldorf/D 1991
Innovationen für die Tischkultur am Beispiel der Kaffee- und Teezubereitung. Innoventa Bremen/D 1992
Mehr Lebensqualität durch innovatives Design. Stilwerk Hamburg/D 1996
Standbau oder Werbebau. Messe Frankfurt Service GmbH, Frankfurt am Main/D 1994

FOTONACHWEIS/PHOTO CREDITS

Umschlag/Cover Joop Greypink, Norbert Latocha, Olaf Gollnek

S./*p. 08/09*	Tassilo von Grolman Design
S./*p. 10/11*	Tassilo von Grolman Design
S./*p. 12/13*	privat
S./*p. 14/15*	Tassilo von Grolman Design
S./*p. 16/17*	Norbert Latocha, Alexander Beck
S./*p. 18/19*	Tassilo von Grolman Design
S./*p. 20/21*	Stockfoto
S./*p. 22/23*	Norbert Latocha, Joop Greypink
S./*p. 24/25*	Tassilo von Grolman Design
S./*p. 26/27*	Joop Greypink, Tassilo von Grolman Design, Norbert Latocha
S./*p. 29*	Joop Greypink
S./*p. 30/31*	Tassilo von Grolman Design
S./*p. 33*	Norbert Latocha
S./*p. 34*	Norbert Latocha
S./*p. 36/37*	Norbert Latocha
S./*p. 38/39*	Stockfoto
S./*p. 40/41*	Gabriele Maßmann, Erbelding Fotografie
S./*p. 42/43*	Tassilo von Grolman Design, Norbert Latocha
S./*p. 44/45*	Norbert Latocha
S./*p. 47*	Bad Pyrmonter Werksfoto
S./*p. 48/49*	Stockfoto
S./*p. 50/51*	Norbert Latocha
S./*p. 52/53*	Sieger Design, Norbert Latocha
S./*p. 54/55*	Norbert Latocha, Joop Greypink
S./*p. 56/57*	Norbert Latocha
S./*p. 58/59*	Stockfoto
S./*p. 60/61*	Privat, Anja Conrad
S./*p. 62/63*	Norbert Latocha
S./*p. 64*	Norbert Latocha
S./*p. 66/67*	Norbert Latocha, Alfi Werksfoto
S./*p. 69*	Anja Conrad
S./*p. 70/71*	Anja Conrad
S./*p. 72/73*	Anja Conrad
S./*p. 74/75*	Anja Conrad
S./*p. 76/77*	Anja Conrad
S./*p. 78/79*	Anja Conrad
S./*p. 80/81*	Anja Conrad
S./*p. 82/83*	Privat, Horst Wackerbarth, Erika Koch
S./*p. 84*	Anja Conrad
S./*p. 86/87*	Stockfoto
S./*p. 88/89*	Norbert Latocha
S./*p. 90/91*	Norbert Latocha
S./*p. 92*	Joop Greypink
S./*p. 94/95*	Norbert Latocha, Tobias Plath
S./*p. 96/97*	Stockfoto
S./*p. 98*	Norbert Latocha
S./*p. 100/101*	Norbert Latocha
S./*p. 102/103*	Lukas Dostal, Alexander Beck
S./*p. 104/105*	Alexander Beck, Norbert Latocha
S./*p. 106/107*	Norbert Latocha, Sieger Design
S./*p. 108/109*	Stockfoto
S./*p. 110/111*	Norbert Latocha
S./*p. 112/113*	Norbert Latocha
S./*p. 115*	Norbert Latocha
S./*p. 116/117*	Norbert Latocha
S./*p. 119*	Alfi Werksfoto
S./*p. 120/121*	Stockfoto
S./*p. 122/123*	Tassilo von Grolman Design
S./*p. 124/125*	Norbert Latocha
S./*p. 126/127*	Norbert Latocha
S./*p. 128*	Norbert Latocha
S./*p. 130/131*	Stockfoto
S./*p. 132*	Norbert Latocha
S./*p. 134/135*	Erika Koch, Heinz Hefele
S./*p. 137*	Norbert Latocha
S./*p. 138/139*	Tassilo von Grolman Design, Joop Greypink
S./*p. 140/141*	Joop Greypink
S./*p. 143*	Joop Greypink